Rooyen T. Mavenyengwa

Molecular, Clinical Microbiology & Diagnosis of Group B Streptococcus

Rooyen T. Mavenyengwa

Molecular, Clinical Microbiology & Diagnosis of Group B Streptococcus

A focus on epidemiology, bacteriology, detection, characterization, prevention, treatment and antibiotic resistance

LAP LAMBERT Academic Publishing

Cover image: www.ingimage.com

Publisher:
LAP LAMBERT Academic Publishing
is a trademark of
Dodo Books Indian Ocean Ltd. and OmniScriptum S.R.L publishing group

120 High Road, East Finchley, London, N2 9ED, United Kingdom
Str. Armeneasca 28/1, office 1, Chisinau MD-2012, Republic of Moldova, Europe
Managing Directors: Ieva Konstantinova, Victoria Ursu
info@omniscriptum.com

Printed at: see last page
ISBN: 978-3-659-81156-2

Molecular, Clinical Microbiology and Laboratory Diagnosis of Group B Streptococcus

Dedication

This book is dedicated to my beloved wife Riya, our son Tidane and daughters Tifane and Tifare and the whole family. Many thanks for the love and support.

A special thanks to Professor Sylvester R. Moyo and Professor Johan A. Maeland for mentoring me in Group B Streptococcus work.

Rooyen Tinago Mavenyengwa 2015
DPhil, MSc, BSc (Hons) Bio. Sci., Grad CE

Table of contents

List of Figures

List of Tables

List of Abbreviations

Alp	Alpha-like proteins
CAMP	Christie Atkins Munch-Peterson
CSF	cerebrospinal fluid
CNA	colistin nalidixic acid agar
CPS	capsular polysaccharide
DNA	deoxyribose nucleic acid
EOD	early onset disease
GBS	group B streptococcus
IAP	intrapartum antibiotic prophylaxis
LOD	late onset disease
MAb	monoclonal antibodies
MLST	Multilocus Sequence Typing
MLVA	Multiple-Locus Variable number tandem repeat Analysis
PCR	Polymerase Chain Reaction
RNA	ribose nucleic acid

Chapter 1

Introduction-Bacteriology and epidemiology of Group B streptococcus

1. Epidemiological background of GBS

1.1 Global GBS

Streptococcus agalactiae or Group B streptococcus (GBS) remains one of the leading causes of neonatal sepsis in many industrialized countries but reports from the developing world infrequently identify this pathogen among newborns with sepsis. Early-onset GBS disease (EOD) occurs within the first week of life and is associated with neonatal sepsis, pneumonia and meningitis. The mortality rate averages at 6.5% for early-onset GBS cases but is higher (22.7%) in infected preterm infants. It has been reported that the proportion of sepsis caused by GBS detected in hospitalized newborns ranged from 0 to 5% for India/Pakistan/Southeast Asia, 0 to 24% for North Africa/Middle East, 0 to 30% for Sub-Saharan Africa and 2 to 35% for the Americas/Caribbean **(1)**. Maternal and neonatal GBS colonization rates and disease vary between different countries or different ethnic groups within the same country. For instance, GBS disease occurs substantially more often in African American infants than in other racial groups. Worldwide, colonization by GBS is highly variable among pregnant women, varying from 1.8% to 36%.

Colonization rates among pregnant women vary greatly due to different choice of sampling method, collection site, the culture medium used, the ethnic group, geographical location, immunological factors and the age of the population investigated. A review of studies with adequate techniques produced regional estimates which were 26% in the US, 12% in India and Pakistan, 19% in Asia, 19% in sub-Saharan Africa, 14% in Central and South America and 22% in the Middle East and North Africa **(2)**. The prevalence of maternal colonization in Scandinavian countries has been reported to range from 21% to 35% **(3, 4)**. In the UK, a mean colonization rate of 14% has been reported. Less is known about the pathogenesis, natural history and transmission dynamics of the organism compared to *S. pyogenes* and *S. pneumoniae*.

1.1.2 GBS in industrialized countries

A systematic review to determine the prevalence of maternal GBS colonization undertaken recently in Europe for the period 1996-2006 **(5)** showed that the regional carriage rates were as follows: Eastern Europe 19.7-29.3%, Western Europe 11-21%, Scandinavia 24.3-36%, and Southern Europe 6.5-32%. GBS serotypes III, II and Ia were the most frequently identified serotypes. Maternal GBS colonization rates in different European countries appear to be similar to those reported in non-European industrialized countries, such as the US (10-30%) **(6)**, Canada (11-19.5%) and New Zealand (20%). The incidence of GBS neonatal disease ranges from 0.5 to 2 per 1000 live births in Europe. In the USA, neonatal disease caused by *S. agalactiae* occurs in approximately 0.69 per 1000 live births and has an estimated mortality rate in neonates of between 4% and 6 %.

1.1.3 GBS in developing countries

In Africa neonatal mortality (first 28 days of life) rate from all causes was about 42 per 1 000 live births between 1995 and 2000; most of which occur in the first week of life, mostly on the first day. During the same period the neonatal mortality rate was estimated 6 per 1000 in Europe and 4 per 1000 in North Americas. The rate was 24 per 1,000 births in Zimbabwe prior to 2005. Neonatal mortality in Asia, Latin America and the Caribbean is about 34, 17 and 19 per 1 000 live births respectively. It is however generally assumed that neonatal mortality in developing countries is under-reported by at least 20%. An estimation of causes of 4 million neonatal deaths for the year 2000 showed 26% was due to sepsis and pneumonia. A proportion of these deaths is probably due to GBS infection. GBS has infrequently been reported in the developing world as a cause of neonatal sepsis despite maternal rectovaginal carriage rates of GBS being similar to those recorded in developed countries including populations in tropical Africa **(2)**. In some African studies the incidence of neonatal sepsis due to GBS has been reported to be low or absent. Other studies from South Africa and Malawi suggest that GBS may be emerging as an important cause of neonatal sepsis in southern Africa. The high carriage rate of this organism in the genital tract of women in Africa has not been associated with high confirmed cases of GBS neonatal sepsis.

1.1.4 GBS: a cause of early onset neonatal sepsis in sub-Saharan Africa

The close relationship between mothers and their infants results in them sharing risk factors and causes of infectious diseases. Data on neonatal morbidity due to bacterial infections in sub-Saharan Africa are very limited. Table1 summarizes some studies providing data on early onset neonatal bacterial sepsis highlighting the role of *E. coli* and *S. agalactiae*. Studies have identified *S. agalactiae* as the most common Gram-positive organism and *E. coli* as the most common Gram-negative organism isolated in neonates younger than 7 days. The data in Table 1 is mainly based on single-site referral hospitals and organisms were isolated from blood and occasionally from CSF.

Table 1.1 GBS and *E. coli* as causes of early onset neonatal sepsis in sub-Saharan Africa

Author Period	(Ref.)	Location	Incidence per 1000 live births (Case fatality rate)	Aetiology
Gray *et al* 2004-2005	(7)	Malawi	0.92* (38%)	*S. agalactiae*-22%[1]*
Berkley *et al* 1998-2002	(8)	Kenya	5.46 (26%)	*E. coli*-19%[1], *S agalactiae*-9%[4]
English *et al* 1999-2001	(9)	Kenya	(18%)	*E. coli*-20%[2], *S. agalactiae*-15%[3]
Milledge *et al* 1996-2001	(10)	Malawi	(48%)	*S. agalactiae*-16%[1], *E. coli*-11%[3]
Madhi *et al* 1997-1999	(11)	South Africa	2.06* (19.8%)	*S. agalactiae*-63%[1]
Mulholland *et al* 1990-1991	(12)	The Gambia	4.42 (42%)	*E. coli*-8%[4], *S. agalactiae*-3%[6],
Nathoo *et al* 1987	(13)	Zimbabwe	21 (28.5%)	*S. agalactiae*-12%[3], *E. coli*-5%[8]

Attributed to GBS alone

*[1-8]In descending order of prevalence; other causative organisms are not listed in the table.

1.2 Group B streptococcus
1.2.1 History and bacteriology
Streptococcus agalactiae is a facultative anaerobe which grows in short chains. It is Gram positive with an ultrastructure similar to other Gram positive cocci. The organism is classified into Lancefield group B by the constituents of its group-

specific cell wall polysaccharide antigen **(14)**. It has a characteristic colonial morphology with a narrow zone of β-haemolysis surrounding the colonies on blood agar plates (**Figure 1.1**). Occasional strains (approximately 1%) are α-haemolytic or nonhaemolytic. Colonies grown on sheep blood agar are 3 to 4 mm in diameter, are grey-white, flat and mucoid. The type specific capsular polysaccharides (CPS) are the basis for classification into serotypes. About 4 to 7% of the strains are nontypable. Antigenic differentiation of GBS strains into serotypes is useful in epidemiological, pathogenetic, immunological, vaccine and taxonomic respects.

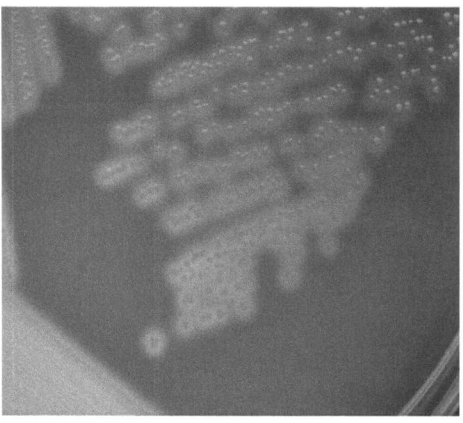

Figure 1.1 *Group B streptococcus colonies on blood agar showing β-haemolysis.*

GBS was initially isolated and described as an animal pathogen causing bovine mastitis. This disease gave the bacterium its name. It was not demonstrated to be a human pathogen until 1938. Neonatal disease, though, was rarely reported. Later during the 1960s numerous reports linked neonatal infections with this organism, and by the 1970s, GBS had become recognized as a leading cause of neonatal infections in much of the developed world. When neonatal GBS infections appeared in the 1970s, up to 50% of patients died. The mortality rate in the 1980s was in the range of 10–15%. During the 1990s, the case-fatality rate of early-onset and late-onset disease was further reduced to approximately 4% because of improvements in neonatal care.

1.2.2 General genomic features of GBS

GBS has a circular genome of around 2.2 million bp with a low G+C content of approximately 35%, typical for streptococci **(15)**. The GBS genome consists of a circular chromosome consisting of a core genome of about 1800 genes shared by all isolates, plus a dispensable genome consisting of partially shared and strain-specific genes. The complete genome has been sequenced for three invasive strains NEM316, 2603V/R and A909 and are just over 2 million base pairs long. Draft sequences have also been obtained from five additional strains: COH1, 515, CJB111, H36B and 18RS21. The *S. agalactiae* genome has a stable backbone and 11–14 interspersed strain variable gene clusters, termed genetic islands. Several known virulence genes are located within these islands. The islands contain known and putative virulence genes, mostly encoding surface proteins and genes related to mobile genetic elements. **Figure 1.2** shows the position and orientation of genes in the GBS strain NEM316.

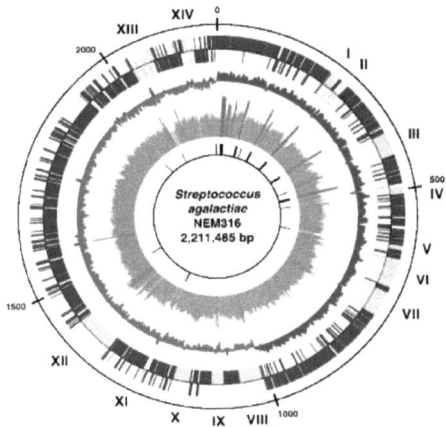

Figure 1.2 *General features of the GBS genome as exemplified by S. agalactiae strain NEM316. The different circles depict the surface protein coding genes, their G/C content and the 14 islands numbered from I to XIV (Source* **(16)***).*

The organizational map of the GBS CPS operon is based on whole genome comparative analysis. The complete DNA sequences of several bacterial CPS and exopolysaccharide synthesis operons have shown that regulatory and structural genes are remarkably well conserved across species. The typical arrangement begins with regulatory determinants, then genes coding for control of polymer chain length and export, followed by structural genes for repeat unit assembly and polymerization, and finally additional genes involved in subunit transport and/or activated monosaccharide biosynthesis (17).

A unique feature of the *S. agalactiae* genome was the identification of 945 genes clustered in 200 regions dispersed around the chromosome. They range in size from 1–77 genes, which can be divided into two groups. The majority of known or putative virulence genes of *S. agalactiae* were found within these regions which may be defined as pathogenicity islands. Based on sequence analysis, 30 open reading frames were predicted in NEM316 encoding putative surface proteins bearing a cell wall sorting signal motif (LPXTG) (18).

1.2.3 GBS antigens
1.2.3.1 Group B streptococcus carbohydrate antigen
Lancefield used a precipitation system for differentiating among types of streptococci into groups according to carbohydrate-containing antigens in their cell walls. Although group B carbohydrate (GBC) appears to be a proinflammatory component, it is difficult to interpret its immunobiological role due to the difficulty in purifying it away from the cell-wall. Passively administered antibodies to group B carbohydrate are not protective against infection in mice. In humans, high levels of maternal antibodies to group B carbohydrate do not confer protection to neonates. The overlying capsule may block access of opsonizing antibody to group B carbohydrate. Group B carbohydrate is the major teichoic acid-like polymer in the cell wall of *S. agalactiae*.

1.2.3.2 GBS capsular polysaccharides
Surface-associated polysaccharides are common features of both Gram-positive and Gram-negative bacteria. GBS isolates have capsular polysaccharides with varied repeating unit structures composed of glucose, galactose, N-acetylglucosamine, and N-acetylneuraminic acid. An analysis of the capsule serotypes revealed two basic motifs, a disaccharide backbone and a variable trisaccharide side chain (19). The repeat unit structures are similar in their constituent monosaccharide compositions and certain structural motifs, yet they differ sufficiently to be antigenically distinct. There are ten different serotypes which have been described namely Ia, Ib and II to IX and almost all isolates can be classified into any of the ten CPS types except for a few that are non-typeable. Lancefield capillary precipitin reaction, latex agglutination and fluorescent antibody test among other methods, have been used for GBS capsular typing. Molecular capsular typing techniques have been elaborated and shown to be reproducible, specific and easy to perform.

Different serotypes cause infections preferably in distinct patient groups. Serotype III causes approximately 37% of early-onset and 67% of late-onset neonatal GBS sepsis compared with 13% and 5%, respectively, caused by serotype V. It is the predominant serotype causing late-onset meningitis (20). Maternofetal and neonatal infections are more likely caused by serotype III sequence type (ST) 17 strains and serotype III restriction digest pattern (RDP) type 3. The genetic elements which determine the propensity of serotype III GBS to cause late-onset sepsis and meningitis have not been fully elucidated. Serotype V causes 29% of invasive GBS infections in non-pregnant adults. It is important however to note that any of the CPS types may cause infection in humans. The CPS elicits type specific antibodies against GBS, which protect against invasive infections but are poorly immunogenic in humans on their own.

1.2.3.3 GBS surface-anchored proteins

Present on most GBS strains is a surface-associated protein antigen which belongs to a certain family of GBS proteins, of which the alpha C protein (Cα) is the prototype **(21)**. The original C protein was shown to be composed of two unrelated protein components, the trypsin-resistant α-protein and the trypsin-sensitive ß-protein **(22)**. The letter "c" historically comes from the designation given by Wilkinson and Eagon when CPS type I**c** was thought to exist as a serotype instead of a protein. The study of surface protein antigens of GBS is important for understanding of the pathogenesis and epidemiology of infection, and several of these antigens have been proposed as components of GBS conjugate vaccines as they induce protective immunity in animal models and are necessary for attachment to the epithelial surface.

Besides the GBS surface-anchored proteins Cα **(22)**, encoded by *bca* and Cβ encoded by *bac*, there are also the classical R proteins R1, R3, and R4. The letter "R" was originally given relating to trypsin **r**esistance. The surface protein called Rib encoded by the gene *rib* (195), seems to be identical to the classical R4 surface protein (21, 190). The alpha-like proteins Alp2 encoded by *alp2* **(23)** and Alp3 encoded by *alp3* **(23)**, may be variants of the classical R1 surface protein. The proteins Cα, Alp1(epsilon), Rib, Alp2, Alp3 and Alp4 belong to the surface protein family called "alpha-like proteins" characterized by similarity in primary structure, with up to 100% homology for some of the surface protein stretches **(23)**, and by their generation of ladder-like patterns on Western blots. The patterns are probably due to large and identical repeat units which vary in number from strain to strain. The surface protein Cβ is not included among the Alps. **Table 1.2** summarizes the C and R surface proteins. **Figure 1.3** shows a diagrammatic presentation of GBS antigens.

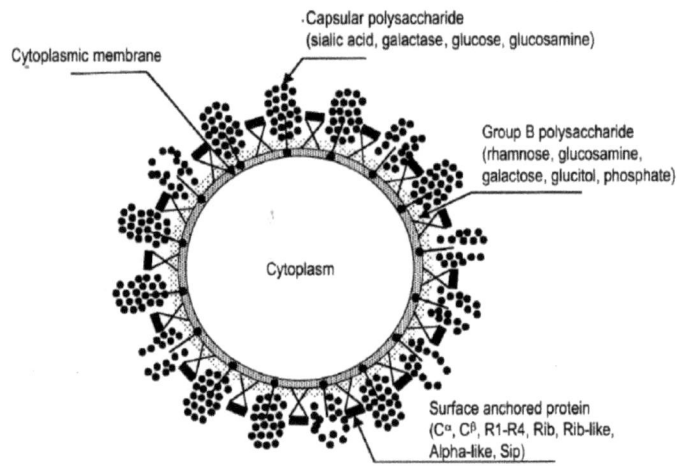

Cytoplasmic membrane

Capsular polysaccharide
(sialic acid, galactase, glucose, glucosamine)

Group B polysaccharide
(rhamnose, glucosamine,
galactose, glucitol, phosphate)

Cytoplasm

Surface anchored protein
(Cᵅ, Cᵝ, R1-R4, Rib, Rib-like,
Alpha-like, Sip)

Figure 1.3 *Diagrammatic presentation of GBS surface antigens (Source Moyo SR, Doctoral thesis, NTNU, 2002).*

Table 1.2 Nomenclature and characteristics of GBS C and R surface proteins.

Designation	Alternate designation	Gene	Most CPS associated	Trypsin susceptibility	Antiserum protective in animal studies	Laddering pattern in Western blot
C proteins						
beta (β)	c beta	*bac*	Ib	S	yes	no
alpha (α)	Alpha-C	*bca*	Ib	R	yes	yes
Alp1(epsilon)	Alp5	*alp1 (epsilon)*	Ia	R	yes	yes
R proteins						
R1	Alp2	*alp2*	III	R	yes	yes
R1	Alp3, R28	*alp3*	V, VIII	R	yes	yes
R3	?	?	V	partially R	yes	yes
R4	Rib, r4	*rib, r4*	III	R	yes	yes
R5	BPS	*sar5*	NT			
Z proteins						
Z1		unknown	V	R	unknown	yes
Z2		unknown	V	R	unknown	yes

The surface proteins are encoded by stable mosaic genes, generated by a recombination of modules at the same chromosomal locus **(23).** The different Alp

8

proteins are encoded by allelic genes, implying that a strain of *S. agalactiae* usually expresses only one member of the family **(23)**. Allelic genes are variant genes that almost always alternatively occupy the same locus in the genome. Horizontal transfer of genetic elements between strains followed by recombinational events has been advocated as an explanation of the structural relatedness and mosaicism of these surface proteins **(23)**. The surface proteins may be important virulence factors in GBS, and they elicit antibodies which are protective in animal models. They hold promise as components in a vaccine based only on surface proteins or as carriers in capsular polysaccharide conjugate vaccines. The sequence organization in Alp family surface proteins reveals a nonrepeated N-terminal region, a repeat region, a wall-anchoring region with an LPXTG motif, a short hydrophobic region that may span the cellular membrane, and a charged tail. Figure 4 shows five major described mechanisms for displaying surface proteins at the surface of Gram-positive bacteria. The anchoring of GBS Alp family proteins is most likely similar to that of Protein A in *S. aureus* shown on the extreme left in **Figure 1.4**.

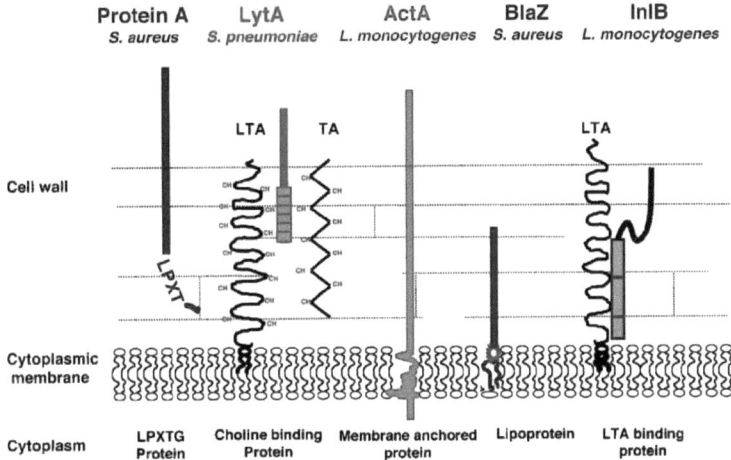

Figure 1.4 *Surface display of major types of surface proteins in Gram-positive bacteria (Source **(24)**).*

1.2.3.3.1 Other surface proteins of *S. agalactiae*

Among the less well studied GBS membrane proteins are R3, R5 and the most recently described Z proteins Z1 and Z2. These proteins were found to be present in a

9

high proportion of GBS strains from pregnant women from Zimbabwe, but less common in clinical isolates from Norway **(25, 26)**. Several other described GBS surface proteins some of which induce protective antibodies are summarised in **Table 1.3** and **Figure 1.5**.

Table 1.3 Other surface proteins of *S. agalactiae*: properties and function.

Protein	(Ref.)	Properties	Function
Streptococcal C5a peptidase	**(27)**	Surface-localized serine protease.	Inactivates human C5a.
Laminin-binding surface protein (Lmb)	**(28)**	Surface-exposed lipoprotein.	Plays a role in colonization and invasion.
Fibrinogen-binding protein (FbsA)	**(29)**	Elicits aggregation of platelets.	Promotes binding of fibrinogen to GBS.
Surface immunogenic protein (Sip)	**(4)**	Present in 9 serotypes.	Elicits protective immunity.
Cell-surface-associated protein (CspA)	**(30)**	Surface-associated protease with an LPXTG motif.	Cleaves fibrinogen.
Glutamine synthetase	**(31)**	Shows sequence similarity to other surface proteins.	Possibly enhances virulence and cell wall synthesis.
Surface protein of group B streptococcus 1 (Spb1)	**(32)**	Has homology to adhesins identified in other bacteria.	Mediates internalization of virulent GBS.
Protein X	**(33)**	Frequently associated with mastitis of dairy cows.	A target of opsonins, could be a protective antigen
Group B secreted protein (Bsp)	**(34)**	Identified in type III strain.	Speculated to control the shape of GBS.
Fbs	**(35)**	Immunologically unrelated to other GBS surface proteins.	Confers protective immunity.
Pho2-2 and Pho3-1	**(36)**	Surface exposed.	Have potential as vaccine candidates.

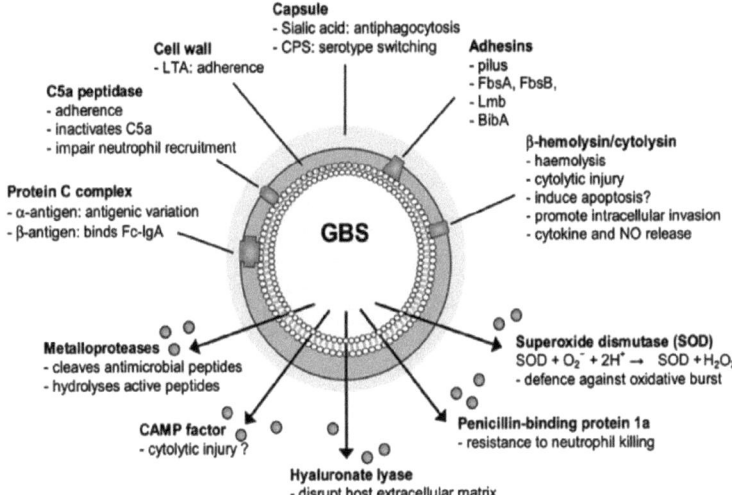

Capsule
- Sialic acid: antiphagocytosis
- CPS: serotype switching

Cell wall
- LTA: adherence

C5a peptidase
- adherence
- inactivates C5a
- impair neutrophil recruitment

Protein C complex
- α-antigen: antigenic variation
- β-antigen: binds Fc-IgA

GBS

Adhesins
- pilus
- FbsA, FbsB,
- Lmb
- BibA

β-hemolysin/cytolysin
- haemolysis
- cytolytic injury
- induce apoptosis?
- promote intracellular invasion
- cytokine and NO release

Metalloproteases
- cleaves antimicrobial peptides
- hydrolyses active peptides

Superoxide dismutase (SOD)
$SOD + O_2^- + 2H^+ \rightarrow SOD + H_2O_2$
- defence against oxidative burst

CAMP factor
- cytolytic injury ?

Penicillin-binding protein 1a
- resistance to neutrophil killing

Hyaluronate lyase
- disrupt host extracellular matrix

Figure 1.5 *Schematic overview of GBS factors involved in adhesion, invasion, avoidance of immune clearance and virulence (Source 15)).*
The list of bacterial features shown is not exhaustive and additional findings are being continuously reported in literature. LTA: lipoteichoic acid; CPS: capsular polysaccharide antigen, Fbs: fibrinogen-binding protein; Lmb: laminin-binding protein; BibA: B Streptococcus immunogenic bacterial adhesion

1.2.3.3.2 Immunological cross-reactivity ladder-forming proteins

There appear to be some cross-reactivity among different alpha-like proteins. **Figure 1.6** illustrates these supposed stretches responsible for cross-reactivity among five members of the alpha-like proteins. It is important to note that no published experimental work has been done to confirm the suppositions discussed which are based on documented sequence homologies.

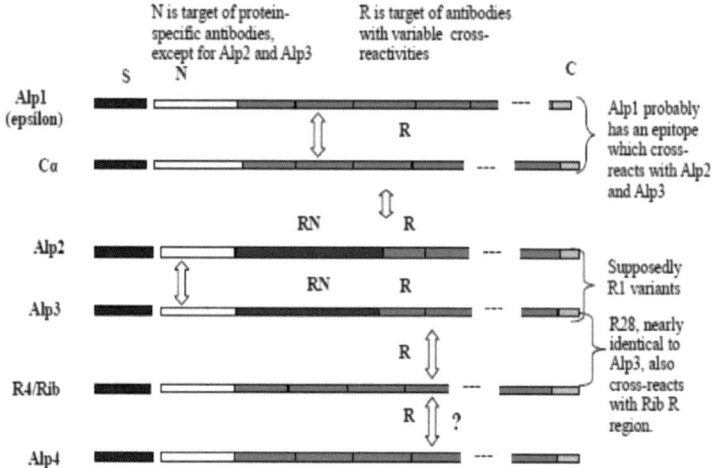

Figure 1.6 Simplified presentation of cross-reactivity among six members of the alpha-like proteins. N: non repeated N terminal region; C: C terminus; R: repeat area; S: signal peptide. Arrow indicate reactivity and cross-reactivity. Compiled in cooperation with Prof Johan A. Maeland, Trondheim, Norway.

The surface protein Alp2 has a site which shows weak cross-reactivity with Cα and possibly with Alp1(epsilon) **(37)**. The cross-reactivity between the surface protein Alp1(epsilon) and Cα imply that Cα and Alp1(epsilon) might be identical immunologically. The Cα monoclonal antibodies (MAb) generated at St. Olav Hospital, Trondheim, Norway, recognizes Cα and epsilon equally well (Maeland, personal communication). Alp1(epsilon) has not been characterized immunologically to the extent comparable to other surface proteins. Purely on the basis of homology of antigenic domains, the N-terminus of Alp2 appears to be immunologically identical to that of Alp3. The antibody binding site was called Alp2/Alp3 common by Maeland *et al* **(37)** and appears to be identical with the classical R1 surface protein specific determinant. Antibodies to this determinant are unable to discriminate between the variants Alp2 and Alp3. The repetitive region of Alp3 cross-reacts completely with that of R4(Rib); an antigenic determinant called Alp3/R4 common. The R4 MAb, until recently considered R4 specific, recognized an epitope localized within the Alp3/R4 common area, meaning that the R4 MAb was not R4-specific **(37)**. Thus Alp3 may not possess a single Alp3-specific antibody-binding site. The R4 N-terminus possesses an antigenic determinant which is R4-specific. Only antibodies against this site can reliably identify R4(Rib), for instance antibody prepared by cross-absorption of an anti-R4 serum with an Alp3 positive strain. Alp3 is considered identical to the GAS R28 surface protein. The surface protein R3 has so far not shown any cross-reactivity when tested using anti-R3 MAb of the IgM isotype. All these cross-reactivities can be explained on the basis of chimeric structures of the Alps. Chimeric sequences are sequences that are made up of portions derived from other sources possibly as a result of recombination by corresponding genes. The exact locations of the various cross-reacting epitopes have not been experimentally confirmed.

1.3 Main references

1. **Stoll, B. J.** 1997. The global impact of neonatal infection. Clin Perinatol **24**:1-21.
2. **Stoll, B. J., and A. Schuchat.** 1998. Maternal carriage of group B streptococci in developing countries. Pediatr Infect Dis J **17**:499-503.
3. **Bergseng, H., L. Bevanger, M. Rygg, and K. Bergh.** 2007. Real-time PCR targeting the sip gene for detection of group B Streptococcus colonization in pregnant women at delivery. J Med Microbiol **56**:223-8.
4. **Valkenburg-van den Berg, A. W., A. J. Sprij, P. M. Oostvogel, J. A. Mutsaers, W. B. Renes, F. R. Rosendaal, and P. Joep Dorr.** 2006. Prevalence of colonisation with group B Streptococci in pregnant women of a multi-ethnic population in The Netherlands. Eur J Obstet Gynecol Reprod Biol **124**:178-83.
5. **Martin, D., S. Rioux, E. Gagnon, M. Boyer, J. Hamel, N. Charland, and B. R. Barcaite, E., A. Bartusevicius, R. Tameliene, M. Kliucinskas, L. Maleckiene, and R. Nadisauskiene.** 2008. Prevalence of maternal group B streptococcal colonisation in European countries. Acta Obstet Gynecol Scand **87**:260-71.
6. **Schrag, S., R. Gorwitz, K. Fultz-Butts, and A. Schuchat.** 2002. Prevention of perinatal group B streptococcal disease. Revised guidelines from CDC. MMWR Recomm Rep **51**:1-22.
7. **Gray, K. J., S. L. Bennett, N. French, A. J. Phiri, and S. M. Graham.** 2007. Invasive group B streptococcal infection in infants, Malawi. Emerg Infect Dis **13**:223-9.
8. **Berkley, J. A., B. S. Lowe, I. Mwangi, T. Williams, E. Bauni, S. Mwarumba, C. Ngetsa, M. P. Slack, S. Njenga, C. A. Hart, K. Maitland, M. English, K. Marsh, and J. A. Scott.** 2005. Bacteremia among children admitted to a rural hospital in Kenya. N Engl J Med **352**:39-47.
9. **English, M., M. Ngama, C. Musumba, B. Wamola, J. Bwika, S. Mohammed, M. Ahmed, S. Mwarumba, B. Ouma, K. McHugh, and C. Newton.** 2003. Causes and outcome of young infant admissions to a Kenyan district hospital. Arch Dis Child **88**:438-43.
10. **Milledge, J., J. C. Calis, S. M. Graham, A. Phiri, L. K. Wilson, D. Soko, M. Mbvwinji, A. L. Walsh, S. R. Rogerson, M. E. Molyneux, and E. M. Molyneux.** 2005. Aetiology of neonatal sepsis in Blantyre, Malawi: 1996-2001. Ann Trop Paediatr **25**:101-10.
11. **Madhi, S. A., K. Radebe, H. Crewe-Brown, C. E. Frasch, G. Arakere, M. Mokhachane, and A. Kimura.** 2003. High burden of invasive Streptococcus agalactiae disease in South African infants. Ann Trop Paediatr **23**:15-23.
12. **Mulholland, E. K., O. O. Ogunlesi, R. A. Adegbola, M. Weber, B. E. Sam, A. Palmer, M. J. Manary, O. Secka, M. Aidoo, D. Hazlett, H. Whittle, and B. M. Greenwood.** 1999. Etiology of serious infections in young Gambian infants. Pediatr Infect Dis J **18**:S35-41.

13. **Nathoo, K. J., P. R. Mason, and T. H. Chimbira.** 1990. Neonatal septicaemia in Harare Hospital: aetiology and risk factors. The Puerperal Sepsis Study Group. Cent Afr J Med **36:**150-6.

14. **Lancefield, R. C.** 1933. A serological differentiation of human and other groups of hemolytic streptococci. J Exp Med **57:**571-595.

15. **Tettelin, H., V. Masignani, M. J. Cieslewicz, C. Donati, D. Medini, N. L. Ward, S. V. Angiuoli, J. Crabtree, A. L. Jones, A. S. Durkin, R. T. Deboy, T. M. Davidsen, M. Mora, M. Scarselli, I. Margarit y Ros, J. D. Peterson, C. R. Hauser, J. P. Sundaram, W. C. Nelson, R. Madupu, L. M. Brinkac, R. J. Dodson, M. J. Rosovitz, S. A. Sullivan, S. C. Daugherty, D. H. Haft, J. Selengut, M. L. Gwinn, L. Zhou, N. Zafar, H. Khouri, D. Radune, G. Dimitrov, K. Watkins, K. J. O'Connor, S. Smith, T. R. Utterback, O. White, C. E. Rubens, G. Grandi, L. C. Madoff, D. L. Kasper, J. L. Telford, M. R. Wessels, R. Rappuoli, and C. M. Fraser.** 2005. Genome analysis of multiple pathogenic isolates of Streptococcus agalactiae: implications for the microbial "pan-genome". Proc Natl Acad Sci U S A **102:**13950-5.

16. **Glaser, P., C. Rusniok, C. Buchrieser, F. Chevalier, L. Frangeul, T. Msadek, M. Zouine, E. Couve, L. Lalioui, C. Poyart, P. Trieu-Cuot, and F. Kunst.** 2002. Genome sequence of Streptococcus agalactiae, a pathogen causing invasive neonatal disease. Mol Microbiol **45:**1499-513.

17. **Chaffin, D. O., S. B. Beres, H. H. Yim, and C. E. Rubens.** 2000. The serotype of type Ia and III group B streptococci is determined by the polymerase gene within the polycistronic capsule operon. J Bacteriol **182:**4466-77.

18. **Medini, D., C. Donati, H. Tettelin, V. Masignani, and R. Rappuoli.** 2005. The microbial pan-genome. Curr Opin Genet Dev **15:**589-94.

19. **Cieslewicz, M. J., D. Chaffin, G. Glusman, D. Kasper, A. Madan, S. Rodrigues, J. Fahey, M. R. Wessels, and C. E. Rubens.** 2005. Structural and genetic diversity of group B streptococcus capsular polysaccharides. Infect Immun **73:**3096-103.

20. **Weisner, A. M., A. P. Johnson, T. L. Lamagni, E. Arnold, M. Warner, P. T. Heath, and A. Efstratiou.** 2004. Characterization of group B streptococci recovered from infants with invasive disease in England and Wales. Clin Infect Dis **38:**1203-8.

21. **Wilkinson, H. W., and R. G. Eagon.** 1971. Type-specific antigens of group B type Ic streptococci. Infect Immun **4:**596-604.

22. **Bevanger, L., and J. A. Maeland.** 1979. Complete and incomplete Ibc protein fraction in group B streptococci. Acta Pathol Microbiol Scand B **87B:**51-4.

23. **Lachenauer, C. S., R. Creti, J. L. Michel, and L. C. Madoff.** 2000. Mosaicism in the alpha-like protein genes of group B streptococci. Proc Natl Acad Sci U S A **97:**9630-5.

24. **Cossart, P., and R. Jonquieres.** 2000. Sortase, a universal target for therapeutic agents against gram-positive bacteria? Proc Natl Acad Sci U S A **97:**5013-5.

25. Mavenyengwa, R.T., Maeland, J.A., and Moyo, S.R. (2009). Putative novel surface-exposed Streptococcus agalactiae protein frequently expressed by the group B streptococcus from Zimbabwe. Clin. Vaccine Immunol.16:1302-1308.

26. Maeland, J.A., Radtke, A., Lyng, R.V., and Mavenyengwa, R.T. (2013). Novel Aspects of the Z and R3 Antigens of Streptococcus agalactiae Revealed by Immunological Testing. Clin. Vaccine Immunol. 20: 607-612.

27. Wexler, D. E., D. E. Chenoweth, and P. P. Cleary. 1985. Mechanism of action of the group A streptococcal C5a inactivator. Proc Natl Acad Sci U S A 82:8144-8.

28. Spellerberg, B., E. Rozdzinski, S. Martin, J. Weber-Heynemann, N. Schnitzler, R. Lutticken, and A. Podbielski. 1999. Lmb, a protein with similarities to the LraI adhesin family, mediates attachment of Streptococcus agalactiae to human laminin. Infect Immun 67:871-8.

29. Schubert, A., K. Zakikhany, M. Schreiner, R. Frank, B. Spellerberg, B. J. Eikmanns, and D. J. Reinscheid. 2002. A fibrinogen receptor from group B Streptococcus interacts with fibrinogen by repetitive units with novel ligand binding sites. Mol Microbiol 46:557-69.

30. Harris, T. O., D. W. Shelver, J. F. Bohnsack, and C. E. Rubens. 2003. A novel streptococcal surface protease promotes virulence, resistance to opsonophagocytosis, and cleavage of human fibrinogen. J Clin Invest 111:61-70.

31. Swingle, H. M., R. L. Bucciarelli, and E. M. Ayoub. 1985. Synergy between penicillins and low concentrations of gentamicin in the killing of group B streptococci. J Infect Dis 152:515-20.

32. Adderson, E. E., S. Takahashi, Y. Wang, J. Armstrong, D. V. Miller, and J. F. Bohnsack. 2003. Subtractive hybridization identifies a novel predicted protein mediating epithelial cell invasion by virulent serotype III group B Streptococcus agalactiae. Infect Immun 71:6857-63.

33. Rainard, P., Y. Lautrou, P. Sarradin, and B. Poutrel. 1991. Protein X of Streptococcus agalactiae induces opsonic antibodies in cows. J Clin Microbiol 29:1842-6.

34. Reinscheid, D. J., C. Stosser, K. Ehlert, R. W. Jack, K. Moller, B. J. Eikmanns, and G. S. Chhatwal. 2002. Influence of proteins Bsp and FemH on cell shape and peptidoglycan composition in group B streptococcus. Microbiology 148:3245-54.

35. Areschoug, T., M. Stalhammar-Carlemalm, C. Larsson, and G. Lindahl. 1999. Group B streptococcal surface proteins as targets for protective antibodies: identification of two novel proteins in strains of serotype V. Infect Immun 67:6350-7.

36. Hughes, M. J., R. Wilson, J. C. Moore, J. D. Lane, R. J. Dobson, P. Muckett, Z. Younes, P. Pribul, A. Topping, R. G. Feldman, and J. D. Santangelo. 2003. Novel protein vaccine candidates against Group B streptococcal infection identified using alkaline phosphatase fusions. FEMS Microbiol Lett 222:263-71.

37. **Maeland, J. A., L. Bevanger, and R. V. Lyng.** 2004. Antigenic determinants of alpha-like proteins of Streptococcus agalactiae. Clin Diagn Lab Immunol **11:**1035-9.

Chapter 2

Pathogenesis and clinical manifestations of GBS infections

2. Pathogenesis of neonatal GBS infections and virulence factors of GBS

2.1 Pathogenesis and virulence factors

Several virulence factors of GBS have been identified. These include capsular polysaccharide, GBS pili, HvgA and β-hemolysin **(1).** The role of capsular polysaccharide has been extensively studied for many years and has been documented as the primary factor that enable GBS to survive in the host by preventing activation of complement pathways involved in opsonophagocytosis **(Figure 2.1).** GBS pili and HvgA are important in the adhesion and attachment of GBS to the host cells. β-hemolysin is implicated in the pulmonary and endothelial cell injuries, which contribute to severe pneumonia in infants because of its cytopathic effects on the pulmonary epithelial cells **(2).**

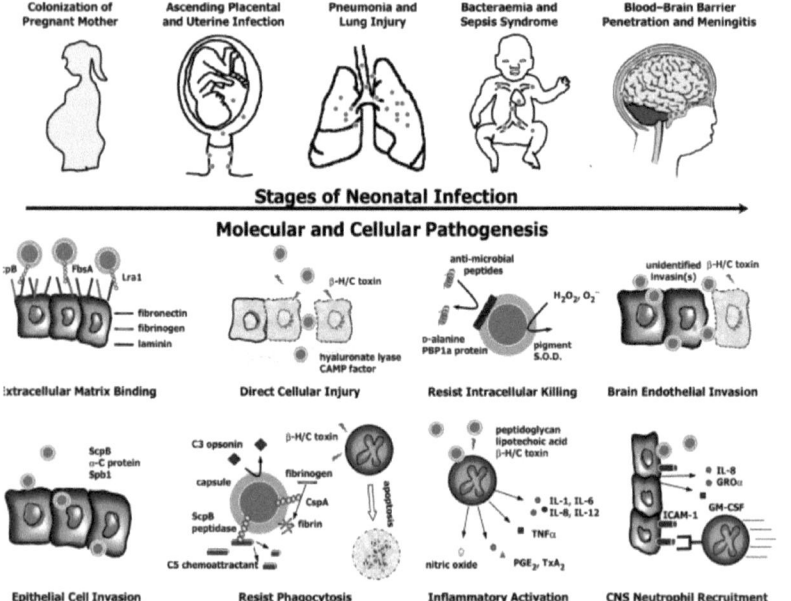

Figure 2.1 *Stages in the molecular and cellular pathogenesis of neonatal GBS infection (Source (3)).*

2.2 GBS disease

2.2.1 Neonatal disease

There are two main clinical forms of neonatal GBS disease; early- and late-onset disease. Lately a third form called infection beyond early infancy has been proposed. Low birth weight, African American ethnicity, prematurity and low apgar scores are examples of neonatal factors found to be associated with GBS disease. Maternal colonization rates correlate with invasive infection rates in neonates as there are several sites from which infection can pass on the unborn baby (**Figure 2.2**).

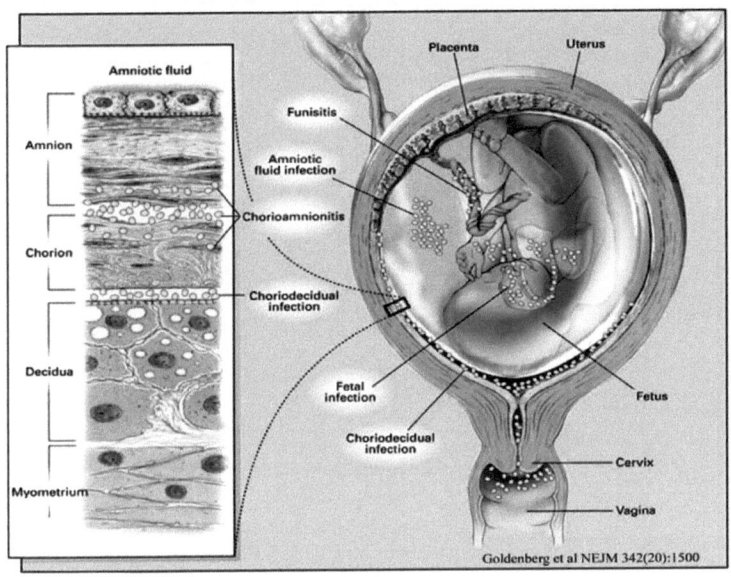

Figure 2.2 *Potential sites of bacterial infection within the uterus. (Source (4)).*

2.2.1.1 Early-onset disease (EOD)

Approximately 80% of neonatal GBS infections called early-onset disease (EOD) occur within the first week of life (**Figure 2.3**). It is vertically transmitted from mother to foetus by exposure to GBS before or during the birth process of which 1%–2% develop early-onset disease. Some early onset infections can occur when the neonate is exposed to GBS during passage through the birth canal, but most early onset infections are probably caused by ascending movement of the organism from the maternal genital area through ruptured membranes into the amniotic fluid. Serotypes III and V cause the majority of infections although any serotype can be isolated from infected infants **(5)**. The presenting signs are lethargy, poor feeding, jaundice, abnormal temperature, grunting respirations, pallor and hypotension. Pneumonia and septicaemia are the most common manifestations. Case fatality rates for EOD are have been reported to range from 10-15% **(5)**. The incidence of early-onset disease is about 10 times higher in premature than in term neonates.

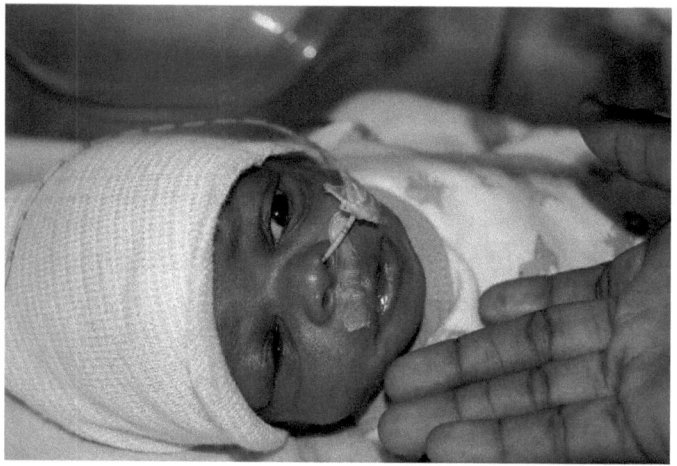

Figure 2.3 Early onset Group B streptococcus in a neonate. (Source http://boybaloskohl27.soup.io/ Accessed 11 November 2015)

2.2.1.2 Late-onset disease (LOD)

Late onset disease (LOD) occurs from 1 week to three months after birth. Serotype III GBS is responsible for approximately 90% of cases of disease **(6)**. Little is known about the pathogenesis but transmission can be either horizontal or vertical. The two most common clinical manifestations are meningitis and bacteraemia. Case fatality rates for late-onset disease are generally lower than for early-onset infections and have been reported to range from 2 to 6% **(6)**. Osteoarticular infections and cellulitis can also occur. The initial signs usually are fever, lethargy, irritability, poor feeding and tachypnoea. Respiratory distress as a presenting feature is less common **(5)**. **Table 2.1** shows a comparison between these two neonatal forms of GBS disease in neonates where the prominence of different serotypes is stated.

Table 2.1 Comparison of neonatal manifestations of GBS disease.

	Risk factors	
	Early-onset disease	**Late-onset disease**
Onset	First week of life, (usually within the first 24 h)	One week to 3 months of age
Clinical presentation	Respiratory distress Pneumonia, Sepsis	Sepsis, Meningitis, Osteoarthritis
Incidence of prematurity	Increased	No change
Maternal obstetrical complications	Frequent (70%)	Uncommon
Transmission	Vertical: acquired in utero or intrapartum	Usually horizontal can also be intrapartum
Predominant serotypes*	Ia, III, V	III, Ia, V
Case fatality rate (%)	10-15	2-6

*In descending order of prevalence. (Source (5)).

2.2.1.3 Outcomes of GBS neonatal disease

In addition to acute illness due to GBS, which is costly to manage, GBS infections in newborns can result in death, disability and recurrence of infection. A substantial proportion of neonates who survive *S. agalactiae* infection suffer from sequelae. Neurological sequelae occur in up to 50% of the survivors of neonatal meningitis and include mental retardation, cortical blindness, deafness, uncontrolled seizures, hydrocephalus, hearing loss, and speech and language delay (6). Neurological sequelae due to GBS are still frequently observed despite major changes in treatment.

2.3 Risk factors for GBS disease in neonates and adults

Some risk factors for EOD and disease in adults have been identified as shown in **Table 2.2**.

Table 2.2 Factors associated with risk for early-onset GBS and disease in adults.

For EOD (7, 8, 9)	For adults (10)
General	**General**
Prolonged membrane rupture > 18 hours	Age of over 60 years
Intrapartum fever	Diabetes
Preterm labour or rapture of membranes <37 weeks gestation	Cancer
	Bed sores
GBS specific	AIDS
Maternal rectovaginal colonization	Long-course corticosteroid therapy
Age under 20 years	Chronic renal disease
Prior birth to an infant with EOD sepsis	Cirrhosis
Heavy colonization	Neurological disorders
Low levels of anti-capsular antibody	
Bacteriuria	
Chorioamnionitis	

2.4 Maternal disease

Vaginal GBS colonization is not associated with disease in healthy non-pregnant women but may cause severe infections in pregnant women (**Figure 2.4**). It is associated with urinary tract infections, bacteremia, amnionitis, endometritis, postpartum wound infections and rarely meningitis (6). Vaginal GBS colonization during pregnancy may increase instances of premature delivery due to GBS ascending infection in women with preterm rupture of membranes. Some of the intra-uterine infections can result in mid-gestation miscarriage episodes even in cases of intact membranes or septic abortion. Caesarian delivery appears to be a prominent risk factor for postpartum endomyometritis. GBS can also cause fever in pregnant women (6).

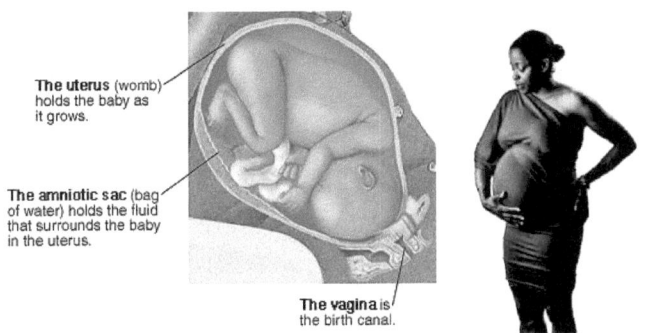

Figure 2.4 Pregnancy can be susceptible to GBS infection (*Source 11, 12)*).

A causal pathway from maternal colonization to infection in the infant is shown in **Figure 2.4**.

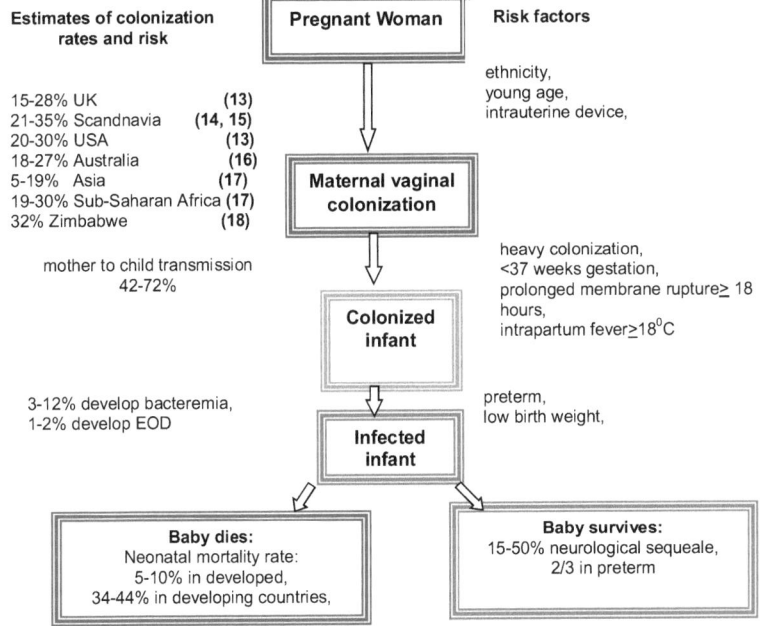

Figure 2.5 Causal pathway of GBS colonization leading to infection and death.

2.5 Spectrum of GBS disease in adults

GBS may occur in a large variety of clinical manifestation in adults. Bacteremia and skin and soft-tissue infections are the most frequently reported manifestation of GBS disease. Other manifestations include Genitourinary infections ,joint and bone infections abdominal infections, endocarditis, Infections of the central nervous system and other miscellaneous infections such as intravascular-device infections, ear, nose and throat infections, endophthalmitis, iatrogenic e.g. post-endoscopy.

2.5.1 Bacteremia

Bacteremia is the presence of bacteria in the bloodstream. This may occur through a wound or injection, or through surgical procedure. Bacteremia may cause no symptoms and resolve without treatment, or it may produce fever and other infections. If the immune system is damaged, septicemia may develop **(19)**.

2.5.2 Sepsis

Sepsis is a severe illness caused by overwhelming infection of the bloodstream by toxin producing bacteria. Neonatal GBS sepsis is characterized by clinical features such as fever, tachypnea and tachycardia. Common sites include the kidneys, liver, skin, and lung. Infection is confirmed by a positive blood culture **(20)**.

2.5.3 Meningitis

Meningitis is the inflammation of the meninges (membranes covering the brain or spinal cord), usually due to infections. Neonatal GBS meningitis occurs by route such as nosocomial infection. It often leads to permanent neurologic sequelae such as cerebral or cranial nerve palsy, epilepsy, mental retardation or hydrocephalus. In adults, GBS meningitis is an important but uncommon manifestation of invasive GBS disease, and may account for up to 4% of all cases of bacterial meningitis. It is almost always associated with anatomical abnormalities contiguous with, or of, the central nervous system, usually as a result of neurosurgery. A small but significant proportion of survivors are left with neurologic sequelae such as permanent hearing loss **(21)**.

2.5.4 Pneumonia

Pneumonia is an infection in one or both lungs. GBS pneumonia is rare and has few unique features. It generally occurs in older adults with diabetes and neurological impairment resulting from conditions such as cerebrovascular disease or dementia. In many cases aspiration is either documented or suspected. Pleural effusions and lung tissue necrosis are rare **(19, 21)**.

2.5.5 Skin and Soft tissue infections
These are the most frequently reported clinical syndromes associated with invasive GBS. These infections most often present as cellulitis, decubitus ulcers, and infected foot ulcers. Many of these patients are diabetic **(19)**. GBS have occasionally been associated with wound and burn infections in 15 non-pregnant adults. Cases of necrotizing fasciitis and toxic shock–like syndrome associated with GBS have been reported.

2.5.6 Bone and joint infections
GBS osteomyelitis most often occurs by contiguous spread or direct inoculation. The bones of the foot are frequently involved; this involvement is linked with vascular insufficiency and overlying ulcers and spreads from adjacent skin and soft-tissue infection. Vertebral osteomyelitis, usually in the lumbosacral area, is another common form of GBS osteoarticular infection **(19, 21)**.

2.5.7 Urinal tract infections
Urinal tract infections (UTI) are more common in older individuals (mean age, 71 years). Many patients with GBS urosepsis are nursing facility residents. In fact, between 5% and 23% of non-pregnant adults with invasive GBS disease are presented with a urinary tract infection. Most patients have significant predisposing conditions, such as diabetes mellitus, prostate disease, an indwelling urinary catheter, and anatomic abnormalities of the urinary tract. The presence of a neurogenic bladder has been associated with significantly increased risk for invasive GBS disease **(21)**. Other conditions to be considered associated with GBS infections include otitis media, endocarditis, neurologic deficit, cellulitis, chorioamnionitis, diabetic foot, line infection, post-partum infections and septic arthritis.

2.6 GBS colonization, transmission
2.6.1 Maternal colonization
The carriage rate of GBS seems to be highest in the rectum suggesting that the lower gastrointestinal tract is an important reservoir. Genital colonization may reflect contamination from the rectum. The organism is present in vaginal flora and inevitably, is transmitted to some babies during labour and delivery, resulting in

colonization **(6)**. It is therefore strategically located to be passed on and cause serious infections in neonates, whose immune response is immature compared to older children and adults **(6)**. Significantly lower colonization rates have been reported for women who are sexually inexperienced, older than 20 years or multiparous. Pregnancy itself does not influence the prevalence of colonization with GBS. The organism may be associated with asymptomatic bacteriuria during pregnancy which is a marker of heavy genital colonization. Colonization prevalence in the vagina and rectum among pregnant women can vary among ethnic groups and geographical locations. Black women have been reported to be colonized at a higher rate as compared to those in other ethnic groups. Maternal GBS colonization has been found to be significantly associated with prolonged labour, premature rupture of membranes, preterm delivery, young age and use of intrauterine device or tampon **(6)**.

2.6.2 Neonatal colonization
In neonates GBS has been isolated from skin, ear, gastric aspirates, throat, umbilicus and rectum **(22)**. They can remain colonized throughout childhood, although little is known about the duration of colonization. Most infants are often colonized at birth by the same serotypes as their mothers. Analysis of mother-infant paired isolates by Western blotting revealed complete concordance with respect to the presence or absence of alpha and beta C protein. Newborns over 48 hours old are most commonly colonized in the throat and rectum **(22)**. At the onset of sexual activity, the genitourinary tract also becomes colonized.

2.6.3 Transmission
Among adults, transmission is hypothesized to occur by the faecal-oral route or by person-to-person direct contact. The colonization of newborns can occur from the mother's vagina, and is acquired during labour and delivery in early-onset disease, although nosocomial, community, and breast milk transmissions have been reported. Transmission however occurs predominantly during labour and approximately 50%-70% of newborns with mothers who are GBS carriers will be colonized during delivery. Prenatal transmission of the organism can occur through both ruptured and intact membranes.

2.6.4 Association between sexual activity and GBS colonization
A high prevalence of GBS carriage among sexually active populations has been described **(23)**. Both men and women can be colonized with GBS as it can be

transmitted sexually. A previous study reported that 36-42% of women visiting sexually transmitted disease clinics had genital GBS colonization **(24)**. Vaginal GBS colonization has been found to be more prevalent among sexually experienced women than among women with no history of sexual intercourse. Pregnant women were found to have an increased risk of genital GBS colonization among those who had frequent sexual intercourse, multiple sex partners and recent acquisition of a new sex partner **(25, 26)**. Vaginal/penile intercourse may cause disturbances of the vaginal ecosystem, which enhances the attachment and growth of GBS in the vagina **(27)**.

2.6.5 Persistent and intermittent maternal GBS colonization dynamics

Maternal GBS carriage in pregnancy has been categorized into chronic (permanent or persistent), transient or intermittent **(6)**. Permanent or persistent carriers are colonized from early pregnancy throughout labour. Others may be positive at some stage of pregnancy but negative at labour i.e. transient colonization **(28)**. New acquisition of genital-tract streptococci may also occur in late pregnancy. The fact that some women are not persistent carriers diminishes the level of prediction of colonization at delivery if tests are done earlier in pregnancy. The transience of colonization raised questions about when women should be screened, and the absence of a rapid test that could be performed during labor raised questions about prenatal screening logistics and strategies for reaching women without prenatal care **(28)**.

The prevention of GBS disease in pregnancy is dependent on the interruption of vertical transmission. Intrapartum antibiotic therapy of GBS carriers has been shown to markedly diminish both neonatal and maternal infection rates. Selective intrapartum approach is effective if there is a good prediction of colonization at the time of delivery. A number of rapid assays for the detection of GBS in labour have been developed but most lack sufficient sensitivity and require heavy GBS colonization to achieve good sensitivity. Longitudinal studies on the predictive value of prenatal GBS cultures have shown varied results and differ in methodology. Whether antenatal cultures would be useful to reliably determine which women should receive intrapartum therapy has represented a controversy in the literature to date.

2.6.6 Invasion or colonization

The ability of certain GBS strains to shift from a colonizing state to a disease-causing state is unclear. Host factors such as genetic predisposition, a compromised immune system or faulty immune response, ethnicity, and behavioural factors have been suggested. Bacterial characteristics such as virulence, adherence capabilities, host evasion mechanisms and inoculum density may contribute to GBS colonization and subsequent disease. A low level of maternal GBS anti-capsular IgG is considered a major risk factor for the development of invasive GBS infections in the neonate.

Newborns with early-onset invasive disease have an increased density of epithelial cell surface receptors specific for GBS. Some GBS strains can replicate more readily in amniotic fluid **(6)** facilitating colonization of the fetus as the first step in disease pathogenesis. Premature birth increases the risk for neonatal GBS infection and the fatality rate is higher among preterm infants. The transplacental transmission of maternal GBS antibodies to the fetus has been found to increase during the third trimester and the amount of maternal antibodies in babies born very prematurely is low. The GBS β-hemolysin/cytolysin induces cytolysis and apoptosis of phagocytes thereby facilitating alveolar epithelial cell injury. If phospholipid-dipalmotyl phosphatidylcholine (DPPC), the major component of human surfactant is increased, the activity of GBS β-hemolysin/cytolysin is reduced. This provides a theoretical rationale for the increased incidence of severe GBS pneumonia in premature, surfactant-deficient neonates. GBS with high levels of lipoteichoic acid (LTA) bound more effectively to buccal cells of fetal rather than adult origin. Production of large quantities of neuraminidase may be a virulence factor in type III GBS **(6)**.

2.7 Main references

1. **Spellerberg, B.** 2000. Pathogenesis of neonatal *Streptococcus agalactiae* infections.
   Microbes Infect **2**:1733-1742.
2. **Nizet, V., Gibson, R. L., Chi, E. Y., Framson, P. E., Hulse, M. and C. E. Rubens.** 1996. Group B streptococcal beta-hemolysin expression is associated with of lung epithelial cells. Infection and Immunity **64**:3818-3826.
3. **Doran K. S., Nizet V.** 2004. Molecular pathogenesis of neonatal group B streptococcal
   infection: no longer in its infancy. Mol Microbiol **54**:23-31.
4. **Goldenberg R. L., J. C. Hauth, and W, W. Andrews.** 2000. Intrauterine infection and preterm delivery. N Engl J Med **342**:1500-7.
5. **Shet, A., and P. Ferrieri.** 2004. Neonatal and maternal group B streptococcal infections: a comprehensive review. Indian J Med Res **120**:141-50.
6. **Edwards, M., Nizet, V, Baker, CJ.** 2006. Group B streptococcal infections, p. 403-464. *In* J. Remington, Klein , JO (ed.), Infectious diseases of the fetus and newborn, 6 ed. Elsevier Saunders, Philadelphia.
7. **Adair, C. E., L. Kowalsky, H. Quon, D. Ma, J. Stoffman, A. McGeer, S. Robertson, M. Mucenski, and H. D. Davies.** 2003. Risk factors for early-onset group B streptococcal disease in neonates: a population-based case-control study. CMAJ **169**:198-203.
8. **Schrag, S., R. Gorwitz, K. Fultz-Butts, and A. Schuchat.** 2002. Prevention of perinatal group B streptococcal disease. Revised guidelines from CDC. MMWR Recomm Rep **51**:1-22.
9. **Schuchat, A., K. Deaver-Robinson, B. D. Plikaytis, K. M. Zangwill, J. Mohle-Boetani, and J. D. Wenger.** 1994. Multistate case-control study of maternal risk factors for neonatal group B streptococcal disease. The Active Surveillance Study Group. Pediatr Infect Dis J **13**:623-9.
10. **Jackson, L. A., R. Hilsdon, M. M. Farley, L. H. Harrison, A. L. Reingold, B. D. Plikaytis, J. D. Wenger, and A. Schuchat.** 1995. Risk factors for group B streptococcal disease in adults. Ann Intern Med **123**:415-20.
11. **Verani, J. R., L. McGee, and S. J. Schrag.** 2010. Prevention of perinatal group B streptococcal disease--revised guidelines from CDC, 2010. MMWR Recomm Rep **59**:1-36.
12. http://www.vanderbilthealth.com/includes/healthtopics/)   Accessed   11

November 2015.

13. Gilbert, R. 2004. Prenatal screening for group B streptococcal infection: gaps in the evidence. Int J Epidemiol 33:2-8.

14. Bergseng, H., L. Bevanger, M. Rygg, and K. Bergh. 2007. Real-time PCR targeting the sip gene for detection of group B Streptococcus colonization in pregnant women at delivery. J Med Microbiol 56:223-8.

15. Valkenburg-van den Berg, A. W., A. J. Sprij, P. M. Oostvogel, J. A. Mutsaers, W. B. Renes, F. R. Rosendaal, and P. Joep Dorr. 2006. Prevalence of colonisation with group B Streptococci in pregnant women of a multi-ethnic population in The Netherlands. Eur J Obstet Gynecol Reprod Biol 124:178-83.

16. Gilbert, G. L., M. C. Hewitt, C. M. Turner, and S. R. Leeder. 2002. Epidemiology and predictive values of risk factors for neonatal group B streptococcal sepsis. Aust N Z J Obstet Gynaecol 42:497-503.

17. Stoll, B. J., and A. Schuchat. 1998. Maternal carriage of group B streptococci in developing countries. Pediatr Infect Dis J 17:499-503.

18. Moyo, S. R., J. Mudzori, S. A. Tswana, and J. A. Maeland. 2000. Prevalence, capsular type distribution, anthropometric and obstetric factors of group B Streptococcus (Streptococcus agalactiae) colonization in pregnancy. Cent Afr J Med 46:115-20.

19. Narayanan S. K. Ossiani M. and Levy C. S. 2006. Streptococcus Group B Infections. eMedicine from WebMD. http://eMedicine-streptococcus-Group-B-infections. Clinical Reference. Page1-14.

20. Humphreys H., Willatts S, and Vincent J. L. 2000. Intensive Care Infections. A practical guide to diagnosis and management in adult patients. WS Saunders. Harcourt Publishers Limited 2000.

21. Farley M.M. 2001. Group B Streptococcal disease in non-pregnant adults. Clin Infect Dis 33:556–561.

22. Hickman, M. E., M. A. Rench, P. Ferrieri, and C. J. Baker. 1999. Changing epidemiology of group B streptococcal colonization. Pediatrics 104:203-9.

23. Manning, S. D., Tallman, P, Baker CJ, Gillespie BW, Marrs CF, Foxman B. 2002. Determinants of co-colonization with group B Streptococcus among heterosexual college partner pairs. Epidemiology 13:533-539.

24. Jackson, D. H., S. M. Hinder, J. Stringer, and C. S. Easmon. 1982. Carriage and transmission of group B streptococci among STD clinic patients. Br J Vener Dis 58:334-7.

25. **Foxman, B., B. W. Gillespie, S. D. Manning, and C. F. Marrs.** 2007. Risk factors for group B streptococcal colonization: potential for different transmission systems by capsular type. Ann Epidemiol **17:**854-62.

26. **Regan, J. A., M. A. Klebanoff, and R. P. Nugent.** 1991. The epidemiology of group B streptococcal colonization in pregnancy. Vaginal Infections and Prematurity Study Group. Obstet Gynecol **77:**604-10.

27. **Meyn, L. A., D. M. Moore, S. L. Hillier, and M. A. Krohn.** 2002. Association of sexual activity with colonization and vaginal acquisition of group B Streptococcus in nonpregnant women. Am J Epidemiol **155:**949-57.

28. **Schrag, S. J.** 2004. The past and future of perinatal group B streptococcal disease prevention. Clin Infect Dis **39:**1136-8.

Chapter 3

Molecular and phenotypic laboratory diagnosis and characterization

3. Laboratory diagnosis of GBS

GBS disease is diagnosed by isolation from a sterile site, primarily blood or cerebrospinal fluid **(1)**. Laboratory diagnosis is done by culture, serological or molecular techniques. The type of infection determines the type of diagnostic samples collected. Other specimens include amniotic fluid, breast milk, urine, rectal, vaginal, ear, nose and umbilical swabs **(2, 3)**. For screening purposes the use of enrichment broth and subsequent culture on solid media results in sample turnaround time of at least two days. Rapid gene-based tests without prior enrichment, theoretically allows screening of pregnant women in labor and give a more accurate diagnosis of colonization than screening in week 35-37.

3.1 Phenotypic techniques

3.1.1 Specimen collection

Swabbing both the lower vagina and anorectal area increases the culture yield substantially compared with sampling of the cervix only while excluding the rectum. The use of appropriate transport media and prompt processing of specimens is important. This sustains the viability of GBS for several days at room temperature; however, the recovery of isolates declines from 1 to 4 days, particularly at high temperatures. Even when appropriate transport media are used, the sensitivity of culture is greatest when the specimen is stored at 4°C before culture and processed within 24 hours of collection **(4)**.

3.1.2 Culture of specimen and medium

The diagnostic standard in pregnant women is culture of vaginal and anal specimens obtained at 35 to 37 weeks of gestation or at delivery when at least one risk factor associated with neonatal infection is present. Isolation of GBS is done from cultured samples in an enriched medium and incubated over 24–48 hours at 35-37°C. Blood agar with and without Lim broth as an enrichment media can be used in identification of GBS. Plating is done on solid media to determine presence of GBS by morphological characteristics. Enriched culture medium is the "Gold Standard" for GBS isolation **(4, 5)**. Antibiotics are included in the selective medium to suppress the overgrowth of other microorganisms that would be part of the flora from the swabbed area.

Important factors that influence the accuracy of detecting GBS maternal colonization are the choice of bacteriological media, body site sampled and timing of sampling. Selective media containing antibiotics are recommended for optimal detection of low levels of GBS colonization of the genital and gastrointestinal tracts (4). They inhibit growth of Gram negative enteric bacilli and other normal flora thereby increasing culture sensitivity for GBS to over 90 per cent.

Selective enrichment broths include; Todd-Hewitt broth supplemented with colistin (10 μg/ml) and nalidixic acid (Lim broth) or Todd-Hewitt broth supplemented with gentamicin (8 μg/ml) and nalidixic acid (15 μg/ml) (TransVag broth). Although TransVag and Lim broth media are often available without blood, the addition of 5% sheep blood can increase the recovery rate of GBS (4). The most widely used medium is Todd-Hewitt broth often supplemented with nalidixic acid, gentamicin, and 5% defibrinated sheep blood. Primary culture onto solid medium has generally been done using colistin-nalidixic acid (CNA) culture medium. Subcultures from broths to blood agar plates are performed, followed by incubation and identification of GBS. The limitations of culture as a method are the time required for a result, the reliance on experience to identify colonies and low sensitivity, making it less useful for detection of GBS infection in labour. The highest culture yield is obtained when both vaginal and anorectal sites are sampled. A positive predictive value of 87% for performing antenatal cultures between 35-36 weeks' gestation has been reported (6).

Some media used relies on production of pigmentation for culture and identification of GBS. However some positive cultures may fail to produce the orange-pigmented colonies that the technique relies on (7). A selective and differential agar medium, known as Granada medium, has been used for the presumptive identification of GBS. Growth of GBS on this medium, under anaerobic conditions, allows the production of orange carotenoid pigmented colonies for 93–98.5% of GBS human isolates (8). Chromogenic media have been introduced to alleviate the detection of GBS in multi-bacterial samples. A color change in the presence of colonies of GBS facilitates detection. The chromogenic or pigmented media works as differential media and indicate growth of GBS by development of a specific color (4). Other media include Columbia CNA agar (CNA), ChromID Strepto B agar (CA) and Northeast Laboratory GBS Screening Medium (NEL-GBS) (9, 10). Use of chromogenic media techniques often reduces time-to-result.

StrepB Carrot Broth, a derivative of Granada medium, is a selective and differential medium for cultivation of GBS. Culture of primary clinical specimens can generate an orange pigment upon overnight incubation and is specific for β-hemolytic strains of *S. agalactiae*. Identification of non-hemolytic strains using these broths alone however may still be challenging **(4, 11)**.

3.1.3 Biochemical testing

There are several presumptive biochemical tests for isolation and identification of GBS. It can be identified by the hippurate hydrolysis reactions and the CAMP test **(Figure 3.1)**. Together with the unique hemolytic reaction (very small zone of lysis 1-2mm around the colony), these two presumptive tests are very accurate in the identification GBS **(12)**. Approximately 96% of GBS isolates are positive for the hippurate hydrolysis test. The CAMP test is presumptuous in a situation whereby a serological method is unavailable **(4)**. Up to 98% of GBS isolates are positive for this test, but some strains of group A streptococci and of Listeria monocytogenes are also CAMP positive It is negative with the pyrrolidonylarylamidase, Voges-Proskauer reaction, hydrolysis of esculin, hydrolysis of starch, production of acid in sorbitol broth deamination of arginine and negative for catalase **(12)**.

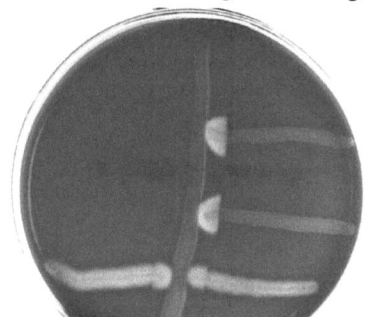
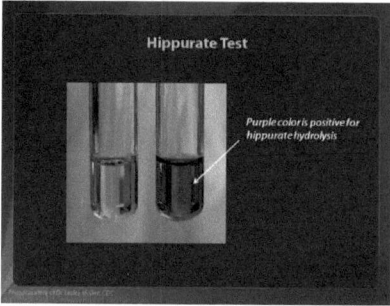

Figure 3.1 CAMP and hippurate hydrolysis test (Source (13)).

3.1.4 Immunological techniques

Serologically, GBS is rapidly identified from cultures using slide agglutination or a capillary precipitation with commercially prepared antisera. Besides grouping, the serotype can be identified by agglutination reactions or precipitation **(14)**. Antisera against polysaccharide antigens can be detected through clumping or precipitation which occurs within a few seconds depending on the group or serotype specific antigen (15, 16). Group B streptococcus strains can also be identified by the production of group B Lancefield antigen (17). Latex agglutination tests and immunoassays that detect the GBS antigen have been developed for rapid detection.

Group B Streptococcus strains are identified by the production of the group B Lancefield antigen (18). Serological typing based upon detection of the group-specific cell wall antigen provides a definitive identification. Rapid diagnostic tests are based on identification of the GBS group-specific polysaccharide antigen from swab specimens and some use enzyme linked immunosorbent (ELISA) technology (19). Although they have good specificity (95%), they tend to have low sensitivity (33 - 65%) which increases with high concentration of bacteria. Various serotyping methods have been used, including immunoprecipitation (20), agglutination (21), counterimmunoelectrophoresis and capillary precipitation (22) and fluorescence microscopy with type-specific antiserum conjugated with fluorescein isothiocyanate (23). A number of commercial assays such as commercial latex agglutination kits, are available and widely used for the rapid detection of GBS in vaginal swab specimens or from culture. Gel filtration, ion-exchange chromatography and Western blotting are also important techniques for analyzing GBS surface proteins. The mouse and the rat are valuable experimental animals for analysis of *S. agalactiae* infections. In passive protection tests, the mouse has been considered the most suitable experimental animal whereas rabbits have been used for generation of antibodies.

3.1.5 MALDI-TOF MS

Matrix-assisted laser desorption ionization–time of flight mass spectrometry (MALDI-TOF MS) is a new technology for species identification based on the protein composition of microbial cells. It is based on different protein profiles characteristic of each microorganism, has been investigated for the identification of GBS and an almost 100% accuracy reported. The most prominent advantages of the technology are speed and low running cost, provided that a quality-controlled database of reference spectra, including all relevant microorganisms, is available. MALDI-TOF MS-based identification also provides much more accurate identification of beta-hemolytic streptococcal species than conventional phenotypic identification methods (24, 25)

3.1.6 Microarrays

DNA microarray is a newly developed technology used for the detection of pathogens and is rapid and sensitive. It consists of four steps: extraction of genomic DNA, amplification of targeted DNA, hybridization of labeled DNA with oligonucleotide probes immobilized on a microarray, and results analysis. A DNA microarray based on the *cpsH* gene in GBS was developed and found to be specific and reproducible. The *cpsH* gene has been identified as a suitable gene to identify GBS serotypes and the GBS DNA microarray, was practicable and a reliable tool for routine identification and serotyping of GBS **(26)**.

3.1.7 Challenges

Non-hemolytic strains can be missed when strains are not screened for the group B antigen. Furthermore, in screening procedures for carrier status, low colonization levels, the presence of other organisms (e.g. in a vaginal culture) and low test sensitivity of the commercial assays pose problems in identifying GBS. Therefore, the detection of GBS isolates and the interpretation of culture results, both from colonized individuals and patients suffering from invasive disease, require a good collaboration between clinicians, microbiologists and laboratory technicians **(27)**. Recent studies have shown that the approach of screening using vaginal and rectal swabs in LIM broth at 35 to 37 weeks misses ~10% of women who become colonized after their screening test, and a substantial number of women convert to negative and would not require prophylaxis **(28)**.Use of certain specimens in antigen testing has its problems. For instance, FDA issued a product alert specifically cautioning against the use of the group B *Streptococcus* antigen kits with urine specimens because of the risk of both false-positive and false-negative results **(29)**.

3.2 Genotypic techniques

Molecular based typing methods for GBS have better sensitivity and low time to result compared to phenotypic methods. Rapid techniques such as DNA probes and nucleic acid amplification tests (NAAT) such as polymerase chain reaction (PCR) have been developed. A real-time based polymerase chain reaction (PCR) such as Strep B assay, GeneXpert GBS assay, BD GeneOhm assay and peptide nucleic acid fluorescence in situ hybridization (PNA-FISH) which are now being employed for the identify GBS directly from sample, enrichment broth, or after subculture **(4, 30-32)**.

Commercial PCR tests designed for this purpose are available, e.g. the BD GeneOhm system (Becton Dickinson, Trondheim, Norway), the GeneXpert system (Cepheid Europe, Maurens-Scopont, France), the probe hybridization system the AccuProbe (San Diego, CA, USA) and a DNA-based probe test developed by Microprobe (Bothell, WA, USA). With rapid processing, these systems may be able to detect the GBS colonization status at delivery and thereby replace screening in weeks 35-37 by antepartum or intrapartum screening. The tests have shown good sensitivity and specificity in some but not all studies (4). Because of their variable performance, the problem of turnaround time under obstetric routine conditions, eventual delays in administration of antibiotics, costs and other unsolved problems, gene-based tests at delivery are for the present regarded as supplemental to screening by culture and risk-based approaches (4, 33).

3.2.1 Polymerase Chain Reaction (PCR)

Recent development of conventional and real-time Polymerase Chain Reaction (PCR) technologies has provided new, reproducible, sensitive, specific and fast detection platforms for GBS bacterial identification and characterization. A number of PCR assays targeting different genes for the specific detection of GBS have been developed. Real time PCRs which detect the CAMP factor gene in GBS (*cfb*) and the surface immunogenic (*sip*) gene have been developed which can be used for screening for GBS colonization by testing vaginorectal specimens (34). PCR has also been designed to identify the ten capsular polysaccharide, surface-anchored proteins and antibiotic resistance genes. Use of PCR and sequencing do not require panels of serotype-specific antisera, which are increasingly difficult to maintain. They provide efficient alternatives to conventional serotyping. Some PCRs used for GBS identification and characterization over the years are listed in **Table 3.1**.

Table 3.1 PCR approaches for detecting and characterizing GBS.

Study Group	Type of PCR	Target gene(s)	Specimen Type
Mawn et al, 1993 **(35)**	Conventional	C (protein)	clinical isolates
Hall et al, 1995 **(36)**	Nested	16S-23S spacer region	csf
Backman et al, 1999; Ahmet et al, 1999 **(37, 38)**	Semi-nested	16S-rRNA	csf
Ke et al, 2000 **(34)**	Real time, Conventional	cfb	vaginal and anal swabs
Kong et al, 2002 **(39)**	Conventional	cps, cfb, 16S-rRNA bca, rib, alp2/alp3, alp4, bac	clinical and reference isolates
Bergseng et al, 2007; Bergh et al, 2004 **(40, 41)**	Real-time	sip	rectal and vaginal swabs
Uhl et al, 2005 **(42)**	Real-time	ptsI	vag/rectal swabs
Creti et al, 2004 **(43)**	Multiplex	bca, rib, alp2/alp3, alp4, epsilon, bac	reference, clinical strains
Kong et al 2005; Zhao et al, 2006 **(44, 45)**	Multiplex/Reverse Line blot	cps, cfb	reference, clinical isolates, bovine strains
Zeng et al, 2006 **(46)**	Conventional, Multiplex	AR-erm, AR related	clinical isolates
Wernecke et al, 2009 **(47)**	Real-time	ssrA	vaginal swabs

3.2.2 Multiplex PCR for GBS characterization

Multiplex PCR can simultaneously amplify several separate regions of DNA such as for the detection of surface protein antigen genes in GBS **(43)**. By allowing direct analysis of the amplicon size, determination of the surface protein antigen genes of alpha C, epsilon, Rib, Alp2, Alp3, and Alp4 proteins, the multiplex PCR assay offers a rapid and simple method of subtyping *Streptococcus agalactiae* based on these genes. The primers for a sequence common to all templates provide a positive control for amplification. The expense of reagents and preparation time is less in multiplex PCR than in systems where several tubes of uniplex PCRs are used **(48)**. Producing some multiplex PCR systems may be as simple as combining two sets of primers for which reaction conditions have been determined separately. However, other multiplex

PCRs must be developed with careful consideration for the regions to be amplified, the relative sizes of the fragments, the dynamics of the primers, and the optimization of PCR technique to accommodate multiple fragments.

A problem with Multiplex assays is where one reaction may outcompete another for DNA polymerase and deoxynucleoside triphosphates resulting in false negatives. Mutiplex PCR however still offers advantage over use of antisera as the latter may not protein specific, where for example, the alpha-C, Alp2 and Alp3 (Alp2/3), and the epsilon proteins cross-react during Western blotting **(39, 49)**. The multiplex PCR-based reverse line blot hybridization assay developed by Borchardt *et al,* **(50)**, uses DNA dot blot hybridization with probes constructed from PCR products has also been used to simultaneously detect antibiotic resistance genes **(46)**. Since it detects the presence of capsule genes rather than capsule expression, it is more specific and reproducible, and is easier to perform than antibody-based serotyping.

3.2.3 DNA hybridization
Probe hybridization for GBS targets specifically the GBS ribosomal RNA. The method has been shown to be suitable to identify GBS from 18h to 24 h cultures in selective enrichment broth with a sensitivity of 94.7-100% and specificity of 96.9-99.5% compared with culture **(51)**. The sensitivity is much lower when incubation is shorter. Thus, available probe hybridization methods are suitable for GBS identification from overnight cultures in selective enrichment broth, but are poorly sensitive for direct detection and identification of GBS from recto vaginal swabs obtained from pregnant women during labour **(51)**.

3.3 GBS typing techniques
Typing of GBS is done to follow the overall composition of GBS strains in a geographic region, for research purposes or for outbreak investigations in humans or in cattle. Typing of GBS starts usually with the determination of its capsular polysaccharide and the strain variable surface proteins. In both cases immunological methods have traditionally been used, but lately, molecular methods testing of the CPS or proteins have become available. A method combining multiplex PCR and reverse line blotting integrates typing of CPS, surface proteins and antibiotic resistance genes in one assay **(44, 46)**. The discriminatory power is comparably low and therefore supplementary methods have been developed. Several molecular typing techniques for defining the clonal relatedness and studying the genetic population structures of GBS strains have been described. These include pulsed-field gel

electrophoresis (PFGE), multilocus enzyme electrophoresis (MEEE) typing multi-locus sequence typing (MLST) among others. Pulsed-field gel electrophoresis is used as a tool for high resolution typing to elucidate epidemiological relations between strains. Multi-locus sequence typing for GBS is appropriate for phylogenetic and epidemiological typing (52). Both PFGE and MLST require sophisticated equipment, well trained personnel and are expensive. Newer more simple and less expensive methods have also been developed such as the MLVA described in this work or analysis of sets of single-nucleotide polymorphisms (SNPs) (53).

3.3.1 Capsular typing

Capsular serotyping is the classic method for typing GBS in epidemiological studies. The CPS antigens can be detected by antibodies raised in rabbits or mice. The classic immunoprecipitation method was described in the 1930s (54). Fluorescent antibody tests (FAT) (56) or agglutination of latex particles (14) have also been used. There is variation in the level of encapsulation among GBS strains (56). Strains have been reported which do not have capsules and these tend to be difficult to type serologically. Horizontal transfer of genes that are relevant for the synthesis and assembly of the CPS may and cause a switching of CPS type (57, 58). This capsule switching may disrupt the phylogenetic relationship of a strain and assign it to a different serotype which may not be in concordance with other typing methods. In-house typing methods of CPS include use of indirect fluorescence antibody test with antibodies raised in rabbits or mice. The Statens Serum Institute in Copenhagen produces a commercially available latex based kit which is widely used by reference laboratories (59).

The sequences of the variable CPS type-specific regions have been made available making it possible to come up with primers to amplify these regions (39). This made it possible to design PCRs for CPS genotyping both as single or multiplex PCRs (44, 60). PCR based typing can identify capsular types in strains found to be non-typable by immunological methods especially in strains from cattle (45).

3.3.2 Surface proteins typing

Strain variable surface proteins have been used for subtyping of CPS types. Some of the best-characterized GBS protein antigens belong to the alpha-like protein (Alp) family, a class of surface proteins characterized by internal long identical tandem repeats. These proteins are named alpha, Alp1, Alp2, Alp3, Alp4, and Rib. The proteins typically appear in combination with certain CPS-types, e.g. III and R4 or Ib and Cα/Cβ unusual CPS/protein combinations may be interesting epidemiologically. Surface protein typing can be done by immunological or molecular methods. Several PCR-based methods for detection of the genes coding for these proteins have been published. Multiplexed PCRs are a convenient approach (43). Most GBS strains usually harbor a surface protein encoding gene although it might not be expressed in some instances (60).

3.3.3 Pulsed-field gel electrophoresis (PFGE)

Pulsed-field gel electrophoresis has been used for typing of GBS since the early 1990s (61). The method is based on the macrorestriction of bacterial DNA by an endonuclease into large fragments. These fragments are then separated by gel electrophoresis where the electric field switches its direction, thereby allowing the large DNA fragments to migrate through the gel. An image of the distribution of bacterial genomic fragments in the gel is interpreted visually or by dedicated software (Figure 3.2). As for many other bacterial species PFGE is a convenient method for typing of GBS. Pulsed-field gel electrophoresis (PFGE), a powerful tool for resolving large DNA molecules, is used for typing bacterial isolates, examination of serotype prevalence and evolutionary divergence, and investigation of outbreaks. PFGE provides increased discrimination compared to PCR-based methods and has been found to be preferable to restriction enzyme digestion with conventional electrophoresis.

This method has however some known drawbacks. It is labour-intensive and requires expensive equipment. It is not easy to obtain comparable images of the same strain in different laboratories with PFGE. However a rapid PFGE method that required 3 days to complete, an improvement over the standard method that required as many as 8 days has been developed (62).This requires rigorously standardized protocols and quality control (63). For PFGE analysis of GBS the restriction enzyme SmaI, which cleaves the DNA chain relatively infrequently, is usually chosen.

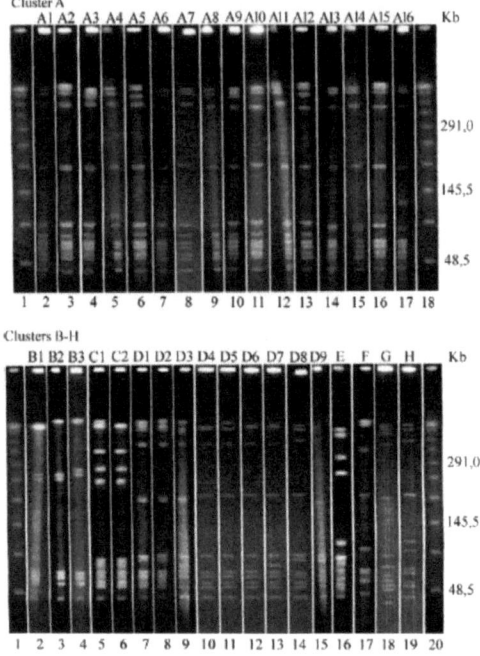

Figure 3.2 *Pulsed-field gel electrophoresis profiles (Source (62)).*

3.3.4 Multi-locus sequence typing (MLST)

A multi-locus sequence typing (MLST) system for GBS has been developed **(52)**. It is based on the sequencing of 400-500 base pairs of the seven housekeeping genes in GBS. Very few mutations in these genes are tolerated and they occur less often than in other parts of the genome. A web-based database is available for comparison and registration of new GBS sequence types (http://pubmlst.org/sagalactiae/). Most of these sequence types belong to clonal complexes (CC) of which four are predominant: CC1, CC17, CC19 and CC23 **(16)**. The complex CC17 has been found in high prevalence among strains from invasive neonatal disease. CC67 is often found in strains from bovine mastitis **(64)**. Clusters made using MLST can be compared with those from Multi-locus variable number of tandem repeats assay (MLVA) **(Figure 3.3)**.

Disadvantages of MLST are its comparable high costs, a sample processing time of 3-4 workdays and the need for a considerable amount of expert hands-on work. Simpler methods which assign presumptive sequence types for the most important types by PCR or SNP analysis have also been developed (53, 65).

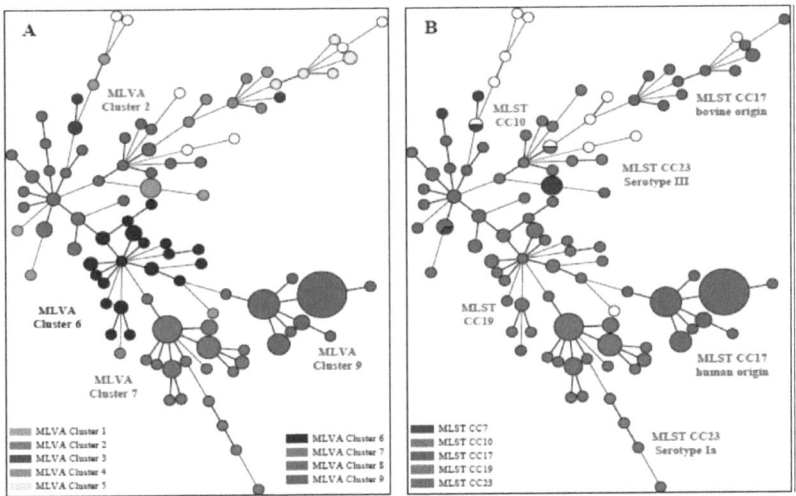

Figure 3.3 *Comparison of MLVA and MLST clustering (Source (66)).*

3.3.5 Multi-locus variable number of tandem repeats assay (MLVA)

Repetitive sequences which may be located both in bacterial genes or intergenic regions are common in prokaryotic genomes (67). These pose a challenge for the genetic copying system of bacteria and deletion or insertion of one or several repeats can occur. As a result, different number of repeats in the bacterial descendant and consequently a different size of the locus occur. A single gene locus with such variations is called a variable number of tandem repeats locus (VNTR). If mispairing occurs it can be repaired by a DNA repair system but the efficiency of these systems is variable (68). Repeated sequences found in intergenic regions are often polymorphic (69). Repeats can also be located in a promoter region of a gene where they can influence the strength of a promoter and cause variation in gene expression (70). A multi locus variable number of tandem repeat analysis (MLVA) scheme for *Streptococcus agalactiae* genotyping has been developed (66). The advantages of MLVA include the ability to analyse bacterial species with high levels of genetic diversity, with six to eight markers being enough for accurate discrimination between strains **(Figure 3.4)**. Highly monomorphic species can still be typed by MLVA, but this requires the use of a larger number of markers.

The method is rapid, cheap and easy genotyping method which generates results suitable for exchange and comparison between different laboratories and for epidemiological surveillance of *S. agalactiae*. The discriminatory power of MLVA may also be increased by adding extra panels of more polymorphic markers or by sequencing repeated sequences displaying internal variability **(67)**.

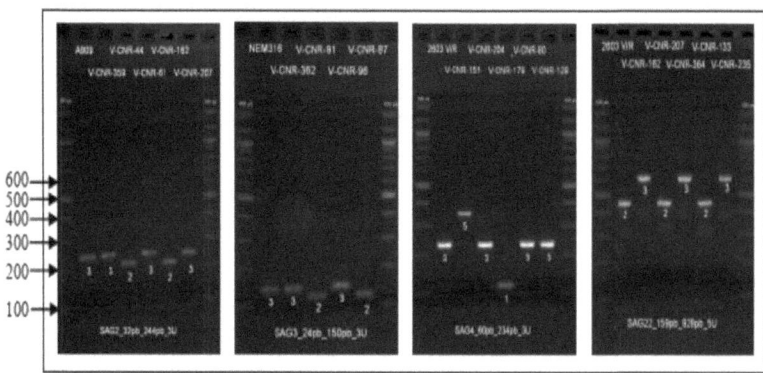

Figure 3.4 *Polymorphism of 4 VNTRs (Source **(66)**).*

3.3.6 Phylogenetic lineage analysis

Subtyping tools have identified specific GBS phylogenetic lineages that are important in neonatal disease. Little is known about the genetic diversity of these lineages or the roles that recombination and selection play in the generation of emergent genotypes. DNA sequencing and PCR-based restriction fragment length polymorphism analysis of several putative virulence genes have been used to examine genetic variation so as to identify gene content differences between genotypes **(71)**. Analysis of the distribution of the various sequence types, clonal complexes and lineages have shown the association of lineage ST-17 serotype III and clonal complex 17 with neonatal invasive disease **(71, 72)**. Data from these different analyses suggest that different GBS gene profiles may be important for disease pathogenesis, as vaccine targets and as markers for the rapid detection of strains causing neonatal disease.

3.3.7 GBS genome sequencing and analysis

Sequencing of the GBS genome has provided valuable information to the understanding of this pathogen and how it cause disease in humans. To date, eight complete sequences and 292 draft GBS genomes have been deposited in the National Centre for Biotechnology Information database (NCBI), and a database called Strepto-DB for comparative genome analysis of group A (GAS) and group B (GBS) streptococci (http://oger.tu-bs.de/strepto_db). The GBS genomes are in the range of 1,800 to 2,160 Kb in size with approx. 1,710 to 2,055 predicted protein coding genes. Some of these genomes were sequenced by the shotgun procedure. There are different sequencing techniques. The oldest and still used has been the Sanger DNA sequencing method. Newer methods have been developed and are referred to as next generation DNA sequencing (NGS) which is combined with advances in bioinformatics.

Genome based phylogenetic analysis in GBS sequenced strains has confirmed the high clonality among GBS lineages mainly containing human strains, and revealed a much higher degree of diversity in the bovine population (73, 74). Two purposes of sequence analysis include finished genome sequence, which represents a complete genome sequence, where the order and accuracy of every base pair have been verified. In contrast, a draft genome sequence represents a collection of contigs of various sizes with unknown order and orientation, that contains sequencing errors and possible misassembles.

3.4 Other typing methods

Other less frequently used methods for typing of GBS include restriction fragment length polymorphism analysis (RFLP) (75), ribotyping (75), multi-locus enzyme electrophoresis (MLEE) (76), random amplification of polymorphic DNA-analysis (RAPD) (77) and amplified cps restriction polymorphism analysis (78).

3.5 Main References

1. **Edwards, M. S., and C. J. Baker.** 2005. Group B Streptococcal infections in elderly adults, Clin Infect Dis. **41**:839–47.
2. **Dore, N., T. D. Bennet, M. Kaliszer, M. Cafferkey, and C. J. Smyth.** 2003. Molecular epidemiology of group B encoding putative virulence factors. Epidemiol Infect **131(2)**:823-833.
3. **Skoff, T. H., M. M. Farley, S. Petit, A. S. Craig, W. Schaffner, K. Gershman, L. H. Harrison, R. Lynfield, J. Mohle-Boetani, S. Zansky, B. A. Albanese, K. Stefonek, E. R. Zell, D. Jackson, T. Thompson, and S. J. Schrag.** 2009. Increasing burden of invasive group B streptococcal disease in nonpregnant adults, 1990–2007. Clin Infect Dis **49**:85-92.
4. **Verani, J. R., L. McGee, and S. J. Schrag.** 2010. Prevention of perinatal group B streptococcal disease--revised guidelines from CDC, 2010. MMWR Recomm Rep **59**:1-36.
5. **Schrag, S. J., E. R. Zell, R. Lynfield, A. Roome, K. E. Arnold, A. S. Craig, L. H. Harrison, A. Reingold, K. Stefonek, G. Smith, M. Gamble, and A. Schuchat.** 2002. A population-based comparison of strategies to prevent early-onset group B streptococcal disease in neonates. N Engl J Med **347**:233-9.
6. **Valkenburg-van den Berg, A. W., R. L. Houtman-Roelofsen, P. M. Oostvogel, F. W. Dekker, P. J. Dörr, and A. J. Sprij.** 2010. Timing of Group B Streptococcus screening in pregnancy: a systematic review. Gynecol Obstet Invest **69**:174–183.
7. **Overman, S. B., D. D. Eley, B. E. Jacobs, and J. A. Ribes.** 2002. Evaluation of methods to increase the sensitivity and timeliness of detection of *Streptococcus agalactiae* in pregnant women. J Clin Microbiol **40**:4329-31.
8. **Rosa-Fraile, M., J. Rodriguez-Granger, M. Cueto-Lopez, A. Sampedro, E. B. Gaye, J. M. Haro, and A. Andreu.** 1999. Use of Granada medium to detect group B streptococcal colonization in pregnant women. J Clin Microbiol **37**:2674–2677.
9. **Aila, N. A. E., I Tency, G. Claeys, B. Saerens, P. Cools, H Verstraelen, M. Temmerman, R. Verhelst, and M. Vaneechoutte.** 2010. Comparison of different sampling techniques and of different culture methods for detection of group B Streptococcus carriage in pregnant women. BMC Infect Dis **10**:285
10. **Carvalho, M. G., R. Richard Facklam, D. Jackson, B Beall, and L. McGee.** 2009. Evaluation of three commercial broth media for pigment detection and

identification of group B streptococci (GBS), *Streptococcus agalactiae*. J. Clin. Microbiol **47(12):**4161-3.

11. **Munson, E., M. Napierala, K. L. Munson, A. Culver, and J. E. Hryciuk.** 2010. Temporal characterization of carrot broth-enhanced real-time PCR as an alternative means for rapid detection of *Streptococcus agalactiae* from prenatal anorectal/vaginal screenings. J Clin Microbiol **48(12):**4495-4500.

12. **Facklam, R.** 2002. What Happened to the Streptococci: Overview of taxonomic and nomenclature changes. Clin Microbiol Rev **15(4):**613-630.

13. https://www.studyblue.com/notes/note/n/micro-ch-89/deck/5852047, Accessed 17 Nov 2015-13

14. **Slotved, H. C., J. Elliott, T. Thompson, and H. B. Konradsen.** 2003. Latex assay for serotyping of group B Streptococcus isolates. J Clin Microbiol **41:**4445-7.

15. **Alfa, M. J., S. Sepehri, P. D. Gagne, M. Helawa, G. Sandhu, and G. K. M. Harding.** 2010. Real-time PCR assay provides reliable assessment of intrapartum carriage of group B *Streptococcus.* J Clin Microbiol **10(1128):**00594-10.

16. **Brochet, M., E. Couve, R. Bercion, J. Sire, and P. Glaser.** 2009. Population structure of human isolates of *Streptococcus agalactiae* from Dakar and Bangui J Clin Microbiol **47(3):**800-803.

17. **Forbes, B. A., D. F. Sahm, and A. S. Weissfield.** Bailey and Scott's *Diagnostic Microbiology*, Eleventh Edition. St. Louis, MO: Mosby, 2002, 1069 pp.

18. **Ruoff, K. L., R. A. Whiley, and D. Beighton.** 1999. Streptococcus, p. 283-296. *In* P. Murray, Baron EJ, Pfaller MA, Tenover FC, Yolken RH (ed.), Manual of clinical microbiology, 7 ed. ASM Press, Washington DC.

19. **Walsh, J. A., and S. Hutchins.** 1989. Group B streptococcal disease: its importance in the developing world and prospect for prevention with vaccines. Pediatr Infect Dis J **8:**271-7.

20. **Baker, C. J., D. K. Goroff, S. L. Alpert, C. Hayes, and W. M. McCormack.** 1976. Comparison of bacteriological methods for the isolation of group of B Streptococcus from vaginal cultures. J Clin Microbiol **4:**46-8.

21. **Hakansson, S., L. G. Burman, J. Henrichsen, and S. E. Holm.** 1992. Novel coagglutination method for serotyping group B streptococci. J Clin Microbiol **30:**3268-9.

22. **Triscott, M. X., and G. H. Davis.** 1979. A comparison of four methods for the serotyping of group B streptococci. Aust J Exp Biol Med Sci **57**:521-7.

23. **Cropp, C. B., R. A. Zimmerman, J. Jelinkova, A. H. Auernheimer, R. A. Bolin, and B. C. Wyrick.** 1974. Serotyping of group B streptococci by slide agglutination fluorescence microscopy, and microimmunodiffusion. J Lab Clin Med **84**:594-603.

24. **Cherkaoui A., S. Emonet J. Fernandez D. Schorderet, and J. Schrenzel.** 2011. Evaluation of matrix-assisted laser desorption ionization-time of flight mass spectrometry for rapid identification of beta-hemolytic streptococci. J Clin Microbiol **49**:3004-5.

25. **Lartigue MF, He´ry-Arnaud G, Haguenoer E, Domelier AS, Schmit PO, van der Mee-Marquet N, P. Lanotte, L. Mereghetti, M. Kostrzewa, and R. Quentin.** 2009. Identification of *Streptococcus agalactiae* isolates from various phylogenetic lineages by matrix-assisted laser desorption ionization-time of flight mass spectrometry. J Clin Microbiol **47**:2284-7.

26. **Wen L., Q. Wang , Y. Li, F. Kong, G. L. Gilbert, B. Cao, L. Wang, and L. Feng.** 2006. Use of a serotype-specific DNA microarray for identification of group B Streptococcus (*Streptococcus agalactiae*). J Clin Microbiol **44(4)**:1447-52.

27. **Sendi, P., L. Johansson, and A. Norrby-Teglund.** 2008. Invasive group B Streptococcal disease in non-pregnant adults : a review with emphasis on skin and soft-tissue infections. Infection **36**:100-111.

28. **Stoll B. J., N. I.Hansen, P. J. Sanchez, R. G. Faix, B. B. Poindexter, K. P. Van Meurs, M. J. Bizzarro, R. N. Goldberg, I. D. Frantz III, E. C. Hale, S. Shankaran, K. Kennedy, W. A. Carlo, K. L. Watterberg, E. F. Bell, M. C. Walsh, K. Schibler, A. R. Laptook, A. L. Shane, S. J. Schrag, A. Das, and R. D. Higgins.** Eunice Kennedy Shriver National Institute of Child Health and Human Development Neonatal Research Network. 2011. Early onset neonatal sepsis: the burden of group B streptococcal and *E. coli* disease continues. Pediatrics **127**:817–826.

29. **U.S. Food and Drug Administration.** 1997. FDA Safety Alert: Risks of Devices for Direct Detection of Group B Streptococcal Antigen. U.S. Food and Drug Administration, Silver Spring, MD.

30. **Ippolito, D. L., W. A. James, , D. Tinnemore, R. R. Huang, M. J. Dehart, J. Williams, M. A. Wingerd, and S. T. Demons.** 2010. Group B Streptococcus serotype prevalence in reproductive-age women at a tertiary care

Military Medical Center relative to global serotype distribution. BMC Infect Dis **10**:336.

31. **Radtke A., B. Lindstedt, J. E. Afset, and K. Bergh.** 2010. Rapid Multiple-Locus Variant-Repeat assay (MLVA) for genotyping of *Streptococcus agalactiae*. J Clin Microbiol **48(7)**:2502–2508.

32. **Peltroche-Llacsahuanga, H., M. Fiandaca, S. von Oy, R. Ltticken, and G. Haase.** 2009. Rapid detection of *Streptococcus agalactiae* from swabs by peptide nucleic acid fluorescence in situ hybridization. J Med Microbiol **59**:179–84.

33. **Spellerberg, B., and C. Brandt.** 2011. Streptococcus, In: Versalovic, J., K. C. Carroll, G. Funke, J. H. Jorgensen, M. L. Landry, D. W. Warnock. (Eds.) Manual of Clinical Microbiology. ASM Press, Washington.

34. **Ke, D., C. Menard, F. J. Picard, M. Boissinot, M. Ouellette, P. H. Roy, and M. G. Bergeron.** 2000. Development of conventional and real-time PCR assays for the rapid detection of group B streptococci. Clin Chem **46**:324-31.

35. **Mawn, J. A., A. J. Simpson, and S. R. Heard.** 1993. Detection of the C protein gene among group B streptococci using PCR. J Clin Pathol **46**:633-6.

36. **Hall, L. M., B. Duke, and G. Urwin.** 1995. An approach to the identification of the pathogens of bacterial meningitis by the polymerase chain reaction. Eur J Clin Microbiol Infect Dis **14**:1090-4.

37. **Backman, A., P. Lantz, P. Radstrom, and P. Olcen.** 1999. Evaluation of an extended diagnostic PCR assay for detection and verification of the common causes of bacterial meningitis in CSF and other biological samples. Mol Cell Probes **13**:49-60.

38. **Ahmet, Z., P. Stanier, D. Harvey, and D. Holt.** 1999. New PCR primers for the sensitive detection and specific identification of group B beta-hemolytic streptococci in cerebrospinal fluid. Mol Cell Probes **13**:349-57.

39. **Kong, F., S. Gowan, D. Martin, G. James, and G. L. Gilbert.** 2002. Molecular profiles of group B streptococcal surface protein antigen genes: relationship to molecular serotypes. J Clin Microbiol **40**:620-6.

40. **Bergseng, H., L. Bevanger, M. Rygg, and K. Bergh.** 2007. Real-time PCR targeting the sip gene for detection of group B Streptococcus colonization in pregnant women at delivery. J Med Microbiol **56**:223-8.

41. **Bergh, K., A. Stoelhaug, K. Loeseth, and L. Bevanger.** 2004. Detection of group B streptococci (GBS) in vaginal swabs using real-time PCR with TaqMan probe hybridization. Indian J Med Res **119 Suppl**:221-3.

42. **Uhl, J. R., E. A. Vetter, K. L. Boldt, B. W. Johnston, K. D. Ramin, M. J. Adams, P. Ferrieri, U. Reischl, and F. R. Cockerill, 3rd.** 2005. Use of the Roche LightCycler Strep B assay for detection of group B Streptococcus from vaginal and rectal swabs. J Clin Microbiol **43**:4046-51.

43. **Creti, R., F. Fabretti, G. Orefici, and C. von Hunolstein.** 2004. Multiplex PCR assay for direct identification of group B streptococcal alpha-protein-like protein genes. J Clin Microbiol **42**:1326-9.

44. **Kong, F., L. Ma, and G. L. Gilbert.** 2005. Simultaneous detection and serotype identification of *Streptococcus agalactiae* using multiplex PCR and reverse line blot hybridization. J Med Microbiol **54**:1133-8

45. **Zhao, Z., F. Kong, G. Martinez, X. Zeng, M. Gottschalk, and G. L. Gilbert.** 2006. Molecular serotype identification of *Streptococcus agalactiae* of bovine origin by multiplex PCR-based reverse line blot (mPCR/RLB) hybridization assay. FEMS Microbiol Lett **263**:236-9.

46. **Zeng, X., F. Kong, H. Wang, A. Darbar, and G. L. Gilbert.** 2006. Simultaneous detection of nine antibiotic resistance-related genes in Streptococcus agalactiae using multiplex PCR and reverse line blot hybridization assay. Antimicrob Agents Chemother **50**:204-9.

47. **Wernecke, M., C. Mullen, V. Sharma, J. Morrison, T. Barry, M. Maher, and T. Smith.** 2009. Evaluation of a novel real-time PCR test based on the ssrA gene for the identification of group B streptococci in vaginal swabs. BMC Infect Dis **9**:148.

48. **Edwards, M. C., and R. A. Gibbs.** 1994. Multiplex PCR: advantages, development, and applications. PCR Methods Appl **3**:S65-75.

49. **Lachenauer, C. S., and L. C. Madoff.** 1996. A protective surface protein from type V group B streptococci shares N-terminal sequence homology with the alpha C protein. Infect Immun **64**:4255-60.

50. **Borchardt, S. M., B. Foxman, D. O. Chaffin, C. E. Rubens, P. A. Tallman, S. D. Manning, C. J. Baker, and C. F. Marrs.** 2004. Comparison of DNA dot blot hybridization and lancefield capillary precipitin methods for group B streptococcal capsular typing. J Clin Microbiol **42**:146-50.

51. **Kircher, S. M., M. P. Meyer, and J. A. Jordan.** 1996. Comparison of a modified DNA hybridization assay with standard culture enrichment for detecting group B streptococci in obstetric patients. J Clin Microbiol **34(2)**:342-4.

52. Jones, N., J. F. Bohnsack, S. Takahashi, K. A. Oliver, M. S. Chan, F. Kunst, P. Glaser, C. Rusniok, D. W. Crook, R. M. Harding, N. Bisharat, and B. G. Spratt. 2003. Multilocus sequence typing system for group B streptococcus. J Clin Microbiol **41**:2530-6.

53. Honsa, E., T. Fricke, A. J. Stephens, D. Ko, F. Kong, G. L. Gilbert, F. Huygens, and P. M. Giffard. 2008. Assignment of *Streptococcus agalactiae* isolates to clonal complexes using a small set of single nucleotide polymorphisms. BMC Microbiol **8**:140.

54. Lancefield, R. C. 1934. A serological differentiation of specific types of bovine hemolytic streptococci (Group B). J Exp Med **59**:441-458.

55. Bevanger, L., and J. A. Maeland. 1977. Type classification of group B streptococci by the fluorescent antibody test. Acta Pathol Microbiol Scand **85B**:357-362.

56. Madoff, L.C., C. L. Paoletti, and D. L. Kasper. 2006. Surface structure of Group B Streptococci important in human immunity, In: Gram-positive pathogens. ASM Press, Washington DC, 169-185.

57. Cieslewicz, M. J., D. Chaffin, G. Glusman, D. Kasper, A. Madan, S. Rodrigues, J. Fahey, M. R. Wessels, and C. E. Rubens. 2005. Structural and genetic diversity of group B streptococcus capsular polysaccharides. Infect Immun **73**:3096-103.

58. Luan, S. L., M. Granlund, M. Sellin, T. Lagergard, B. G. Spratt and M. Norgren. 2005. Multilocus sequence typing of Swedish invasive group B streptococcus isolates indicates a neonatally associated genetic lineage and capsule switching. J Clin Microbiol **43**:3727-3733.

59. Afshar, B., K. Broughton, R. Creti, A. Decheva, M. Hufnagel, P. Kriz, L. Lambertsen, M. Lovgren, P. Melin, G. Orefici, C. Poyart, A. Radtke, J. Rodriguez-Granger, U. B. S. Sorensen, J. Telford, L. Valinsky, and L. Zachariadou. Members of the DEVANI Study Group. A. Efstratiou. 2011. International external quality assurance for laboratory identification and typing of *Streptococcus agalactiae* (Group B Streptococci). J Clin Microbiol **49**:1475-1482.

60. Imperi, M., M. Pataracchia, G. Alfarone, L. Baldassarri, G. Orefici, and R. Creti. 2009. A multiplex PCR assay for the direct identification of the capsular type (Ia to IX) of *Streptococcus agalactiae*. J Microbiol Methods **80**:212-214.

61. Fasola, E., C. Lindahl, and P. Ferrieri. 1993. Molecular analysis of multiple isolates of the major serotypes of group B streptococci. J Clin Microbiol **31**:2616-2620.

62. Corrêa A. B., I. C. Oliveira, T. Pinto, M. Mattos, and L. Benchetrit. 2009. Pulsed-field gel electrophoresis, virulence determinants and antimicrobial susceptibility profiles of type Ia group B streptococci isolated from humans in Brazil. Mem Inst Oswaldo Cruz, **104(4):**599-603.

63. van Belkum, A., P. T. Tassios, L. Dijkshoorn, S. Haeggman, B. Cookson, N. K. Fry, V. Fussing, J. Green, E. Feil, P. Gerner-Smidt, S. Brisse, and M. Struelens. European Society of Clinical Microbiology and Infectious Diseases Study Group on Epidemiological, Markers. 2007. Guidelines for the validation and application of typing methods for use in bacterial epidemiology. Clin Microbiol and Infect **13**:1-46.

64. Sørensen, U. B. S., K. Poulsen, C. Ghezzo, I. Margarit, and M. Kilian. 2010. Emergence and global dissemination of host-specific *Streptococcus agalactiae* clones. mBio 1. mBio **1(3):**e00178-10.

65. Lamy, M. C., S. Dramsi, A. Billoet, H. Reglier-Poupet, A. Tazi, J. Raymond, F. Guerin, E. Couve, F. Kunst, P. Glaser, P. Trieu-Cuot, and Poyart, C. 2006. Rapid detection of the "highly virulent" group B Streptococcus ST-17 clone. Microbes Infect **8**:1714-1722.

66. Haguenoer E., G. Baty, C. Pourcel, M. Lartigue, A. Domelier, A. Rosenau, R. Quentin, L. Mereghetti, and P. Lanotte. 2011. A multi-locus variable number of tandem repeat analysis (MLVA) scheme for *Streptococcus agalactiae* genotyping. BMC Microbiol **11**:171.

67. Lindstedt B. A. 2005. Multiple-locus variable number tandem repeats analysis for genetic fingerprinting of pathogenic bacteria. Electrophoresis **26(13):**2567-82.

68. Caporale, L. H. 2003. Natural selection and the emergence of a mutation phenotype: an update of the evolutionary synthesis considering mechanisms that affect genome variation. Annu Rev Microbiol **57**:467-485.

69. Pourcel, C., and G. Vergnaud. 2011. Strain typing using Multiple-loci "Variable Number of Tandem Repeat" analysis and genetic element CRISPR, In: Persing, D.H. (Ed.) Molecular Microbiology: Diagnostic Principles and Practice, Second Edition. ASM Press, Washington, 179-197.

70. Martin, P., T. van de Ven, N. Mouchel, A. C. Jeffries, D. W. Hood, and E.R. Moxon. 2003. Experimentally revised repertoire of putative contingency loci

in *Neisseria meningitidis* strain MC58: evidence for a novel mechanism of phase variation. Mol Microbiol **50**:245-257.

71. **Springman, A. C., D. W. Lacher, G. Wu, N. Milton, T. S. Whittam, H. D. Davies, and S. D. Manning.** 2009. Selection, recombination, and virulence gene diversity among group B streptococcal genotypes. J Bacteriol **191**:5419-27.

72. **Bisharat, N., N. Jones, D. Marchaim, C. Block, R. M. Harding, P. Yagupsky, T. Peto, and D. W. Crook.** 2005. Population structure of group B streptococcus from a low-incidence region for invasive neonatal disease. Microbiology **151**:1875-81.

73. **Rosini, R., E. Campisi, M. Chiara, H. Tettelin, D. Rinaudo, C. Toniolo, M. Metruccio, S. Guidotti,U. B. S. Sørensen, M. Kilian. DEVANI Consortium. M. Ramirez. R. Janulczyk, C. Donati, G. Grandi, I. Margarit.** 2015. Genomic analysis reveals the molecular basis for capsule loss in the Group B Streptococcus population. PLoS ONE **10(5)**:e0125985.

74. **Tettelin, H., V. Masignani, M. J. Cieslewicz, C. Donati, D. Medini, N. L. Ward, S. V. Angiuoli, J. Crabtree, A. L. Jones, A. S. Durkin, R. T. Deboy, T. M. Davidsen, M. Mora, M. Scarselli, I. Margarit y Ros, J. D. Peterson, C. R. Hauser, J. P. Sundaram, W. C. Nelson, R. Madupu, L. M. Brinkac, R. J. Dodson, M. J. Rosovitz, S. A. Sullivan, S. C. Daugherty, D. H. Haft, J. Selengut, M. L. Gwinn, L. Zhou, N. Zafar, H. Khouri, D. Radune, G. Dimitrov, K. Watkins, K. J. O'Connor, S. Smith, T. R. Utterback, O. White, C. E. Rubens, G. Grandi, L. C. Madoff, D. L. Kasper, J. L. Telford, M. R. Wessels, R. Rappuoli, and C. M. Fraser.** 2005. Genome analysis of multiple pathogenic isolates of Streptococcus agalactiae: implications for the microbial "pan-genome". Proc Natl Acad Sci U S A **102**:13950-5.

75. **Blumberg, H. M., D. S. Stephens, C. Licitra, N. Pigott, R. Facklam, B. Swaminathan, and I. K. Wachsmuth.** 1992. Molecular epidemiology of Group B Streptococcal infections: Use of Restriction Endonuclease Analysis of chromosomal DNA and DNA Restriction Fragment Length Polymorphisms of Ribosomal RNA genes (ribotyping). J Infect Dis **166**:574-579.

76. **Quentin, R., H. Huet, F. Wang, P. Geslin, A. Goudeau, and R. Selander.** 1995. Characterization of *Streptococcus agalactiae* strains by multilocus enzyme genotype and serotype: identification of multiple virulent clone families that cause invasive neonatal disease. J Clin Microbiol **33**:2576-2581.

77. Limansky, A. S., E. G. Sutich, M.C. Guardati, I. E. Toresani, and A. M. Viale. 1998. Genomic diversity among *Streptococcus agalactiae* isolates detected by a degenerate Oligonucleotide-Primed Amplification assay. J Infect Dis **177**:1308-1313.

78. Manning, S. D., D. W. Lacher, H. D. Davies, B. Foxman, B., and T. S. Whittam. 2005. DNA polymorphism and molecular subtyping of the capsular gene cluster of Group B Streptococcus. J Clin Microbiol **43**:6113-6116.

Chapter 4

GBS prevention, treatment and antibiotic resistance

4. GBS prevention and treatment

4.1 Strategies for perinatal GBS prevention

The vertical transmission of GBS from the mother to the child and the risk for early onset disease (EOD) can be prevented by administering intrapartum antibiotic prophylaxis (IAP) or by vaccination. A decision on a prophylactic regimen to follow should be based on the epidemiology of GBS in the population. Currently few countries have implemented intrapartum antibiotic prophylaxis strategies despite a 70% decline in the number of early-onset neonatal GBS cases in the USA since its use (1).

According to the Centre for Disease Control and Prevention (CDC), pregnant women who had a previous baby with GBS disease, urinary tract infection due to GBS, had fever during labor, prolonged or difficult labor, rupture of membranes before 37 weeks of pregnancy, and rupture of membrane 18 hours before delivery, are at higher risk of having a baby with GBS disease. The Centre for Disease Control and Prevention recommends the use of either risk assessment or screening for GBS colonization in pregnant women to identify candidates for intrapartum prophylaxis. Risk assessment is performed at the onset of labor and maternal risk factors are assessed and those considered indicative of the need for prophylaxis identified. The screening approach to prevention of GBS infections in pregnancy has become standard in some countries. This approach requires intrapartum antimicrobial prophylaxis in full term women who are GBS culture positive from the vagina or rectum. These women can be given intravenous antibiotics during labor; starting 4 hours before delivery. These recommendations are based on the fact that administration of antimicrobial agents during labor to women at risk of transmitting GBS to their newborns could prevent invasive disease in the first week of life (2).

A large population-based study conducted during 1998–1999 demonstrated the superiority of culture-based screening over the risk-based approach to prevention of early-onset GBS disease (3). The study found that culture-based screening resulted in the identification of a greater proportion of women at risk for transmitting GBS to their newborns. Furthermore, women with a positive antenatal GBS culture were more likely to receive intrapartum antibiotic prophylaxis than those women with a risk-based indication for chemoprophylaxis. In 2002, CDC's guidelines for GBS prevention were updated to recommend universal culture-based screening to determine which women should receive intrapartum GBS chemoprophylaxis (4). CDC recommended that women with unknown GBS colonization status at the time of delivery be managed according to the presence of intrapartum risk factors (**Figure 4.1**).

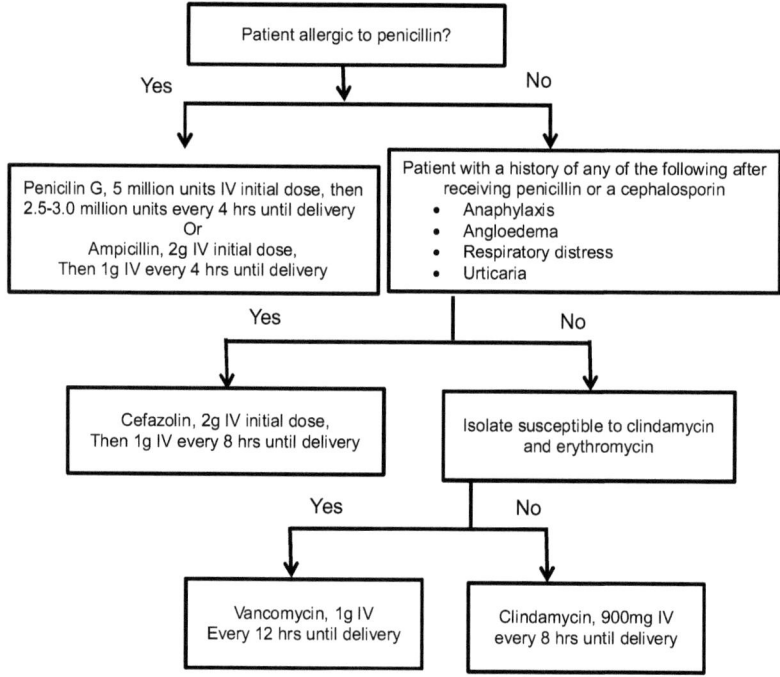

Figure 4.1 *Recommended regimens for intrapartum antibiotic prophylaxis for prevention of early-onset group B streptococcal (GBS) disease (Source (5)).*

In Norway for instance, guidelines for prevention of GBS were issued in 1998 and new guidelines in 2007/2008, recommending IAP to women with risk factors (www.legeforeningen.no). In Sweden there exist routines for regular check-ups of women during pregnancy at antenatal clinics and there is no national program yet but most centers use a risk-based program (www.infpreg.se). Also in other parts of

Europe a risk-based program is generally used (5, 6). The potential for resistance to the commonly used antibiotics during prophylaxis is a concern, especially if there is allergy to penicillin. Antimicrobial resistance to clindamycin and erythromycin by GBS has emerged (7, 8). The main concern for intrapartum chemoprophylaxis is that it does not prevent late-onset disease.

4.1.1 Immunoprophylaxis to prevent GBS disease

Active immunization with a GBS vaccine for prevention of GBS disease has several advantages which include protection of neonates against EOD and late onset disease (LOD) including maternal and adult disease. The vaccine should preferably eliminate maternal carriage of GBS as this reduces neonatal exposure especially if maternal transfer of antibody is inadequate.

Vaccines against GBS were studied early when it was shown that capsular serum antibodies were protective (9) but an effective vaccine is still to be formulated. Before the 90s, several immunization studies with purified capsular polysaccharide (CPS) from the main serotypes were done but were poorly immunogenic in adults (10, 11). To enhance the immunogenicity of CPS, they were conjugated to carrier proteins. This conjugation leads to activation and clonal expansion of carrier-specific T-cells, which led to induction of immunologic memory and better response to the CPS antigens. Polysaccharides have been conjugated to the tetanus toxoid (TT) and immunization studies in animals showed much improved immunogenicity than uncoupled CPS (12, 13). Some studies with coupled CPS vaccines in adults and pregnant women have showed an improved response to the vaccines (14, 15). Correlation between IgG antibodies in maternal and cord sera after immunization indicates efficient transfer of specific antibodies to the fetus (15).

Group B Streptococcus surface proteins elicit protective immunity in animal models as they confer protective immunity (16). Some surface proteins coupled to CPS were immunogenic and at the same time enhanced the immunogenicity of the CPS. Surface proteins can also act as vaccine components alone or in combination. A combined rib and alpha C protein vaccine was shown to protect against a majority of infections with GBS strains in a mouse model (17).

4.1.2 Vaccine development and its future

Rebecca Lancefield detected that protection against GBS infection in mice could be achieved using CPS-specific polyclonal rabbit serum (18). Neonatal resistance to invasive GBS infection is associated with the presence of maternal antibodies (19).

Based on this, maternal immunization has been suggested as the best strategy for the prevention of GBS infection in neonates **(19)**. Capsular polysaccharide-tetanus toxoid conjugate vaccines effective against nine serotypes have been prepared and shown to induce active CPS-specific IgG **(20)**. Clinical trials with conjugate vaccines prepared with purified CPS types Ia, Ib, II, III and V have been shown to be safe and immunogenic **(21)**. A vaccine formulation of GBS types Ia, II, III and V, for instance would be expected to provide protection against over 90 per cent of invasive infant and maternal disease **(22)**. The principal difficulty in developing globally effective GBS vaccines is the existence of different serotypes within different geographical locations; a vaccine suitable for Asian populations might not be suitable for African populations **(23)**. Research and development of a universally effective vaccine based on the GBS surface protein that exploits the recently acquired genomic sequences of GBS strains is continually being done. Potential protective antigens which have been described include unmodified and conjugate polysaccharides and surface proteins, Cα, Cß, Rib protein antigens **(24)**, surface immunogenic protein, Sip **(25)** and C5a peptidase. Epidemiological surveillance of serotype distribution in different populations around the world remains critical for vaccine studies. **Table 4.1** summarizes the various vaccine targets used in GBS vaccine development over the years.

Table 4.1 Summary of GBS vaccine research and development strategies.

Vaccine Target	Approach/(dis)advantage	Period (References)	
Capsular polysaccharide			
CPS alone	Protection tests in mice, immunized with partially purified, type-specific polysaccharides.	1934-38	(18)
	Low immunogenicity of CPS shown.	1970s	(19)
CPS-protein conjugate	Coupling a protein to CPS (Cα and type III), mouse protection study. Improved immunogenicity compared to CPS alone but similar to CPS-TT conjugates.	Late 1990s	(26)
	Recombinant Alp3 as a carrier protein for GBS type III CPS.	2008	(27)
CPS-TT conjugate	Coupling CPS to tetanus toxoid, infection of murine models. Human trials.	1999-2002	(20, 28)
Proteins			
Cβ	Elicited protective immunity in animal models but present in few strains.	1992	(16)
Sip	Expressed by all serotypes, Recombinant Sip protection study in mice.	2000	(25)
C5a peptidase	Protein expressed by all serotypes. Capacity of mouse bone marrow macrophages and human PMNs to kill GBS in vitro when directed against C5a peptidase opsonised GBS.	2000	(29)
	Immunized mice with encapsulated C5a and measuring antibody response by ELISA. Elicited significant immune responses and protection against GBS challenge.	2002	(30)
LmbP	Expressed by most GBS strains	2005	(31)
LrrG	Highly conserved protein. Genomic and immunological assays	2005	(32)
Fusion protein from Rib and Cα	Fusion protein from the N-terminal portions, multigenome analysis and screening, invitro opsonophagocytic activity and mouse protection assays. Highly immunogenic.	2007	(33)
GBS type I pullulanase (SAP)	Anti-SAP functional antibodies may reduce bacterial fitness in vivo and interfere with the capacity of GBS to colonize host tissues.	2008	(34)
Pili	Combination of 3 pili types, genomic and protection assays. Present in all tested GBS strains.	2008	(35)

CPS, capsular polysaccharide; LmbP, laminin binding protein; LrrG, leucine-rich repeat protein; Sip, surface immunogenic protein; TT, tetanus toxoid

4.1.3 Chlorhexidine treatment

Use of chlorhexidine vaginal wipes or douches to reduce sepsis and vertical transmission of GBS has produced contrasting results in different studies. Non-randomized studies have yielded promising results **(36)** but randomized clinical trials have found no protection against early-onset GBS disease or neonatal sepsis. Chlorhexidine vaginal treatment, with or without neonatal wash, was shown to reduce GBS bacterial load but did not have any impact on EOD. Still, a systematic review that included non-randomized studies has suggested that a reduction in maternal and neonatal sepsis may be achieved in developing countries by the use of this method **(37)**.

4.2 Treatment approach in GBS infections

There are several regimens for the treatment of infections with GBS. Antibiotics currently used for management of GBS include penicillin G, ampicillin, vancomycin, cefuroxime, cefotaxime, ceftriaxone, erythromycin, clindamycin, gentamicin, chloramphenicol, ciprofloxacin, and amoxicillin. Penicillin remains the drug of choice for antimicrobial intrapartum prophylaxis and treatment, but allergic women who have a high risk for anaphylaxis could use erythromycin or clindamycin **(2, 38)**. Antibiotic susceptibility testing is necessary before treatment is given. Duration of treatment depends on the clinical syndrome, risk of complications, response to therapy and the age of the patients **(39, 40)**.

4.2.1 Treatment of GBS infection in newborns

Antibiotic treatment depends on the specific area of the body where infection has taken place. Manifestation of signs and symptoms is not specific in neonates. All neonates are therefore treated almost the same once GBS infection is confirmed. Syndromic treatment is often preferred after taking the necessary samples have been taken for initial screening through the Gram stain and culture. Most invasive GBS strains are susceptible to penicillin G and hence the first-line treatment is benzylpenicillin (Penicillin-G) or ampicillin, plus gentamicin. For children with penicillin allergy, a 2nd or 3rd generation cephalosporin (e.g. cefuroxime, cefotaxime or ceftriaxone) may be appropriate depending on the type of allergy, or vancomycin with or without gentamicin. Vancomycin is not suitable as monotherapy if meningitis is present. In the case of septic arthritis, joint must be aspirated and a formal washout in operating room is strongly recommended. Treatment course is 3 to 4 weeks. An infectious-disease specialist should be consulted if polymicrobial infection is suspected **(7, 40, 41)**. Although some studies indicate that 6-7 days therapy might be sufficient for uncomplicated bacteraemia, recommended duration of treatment of GBS infections has been 10-14 days for bacteraemia without focus or with soft tissue infection, 2 to 3 weeks for meningitis or bacterial arthritis and 3 to 4 weeks for osteomyelitis.

4.2.2 Treatment of GBS infection in pregnant women
The most common infections in women are urinary tract infections, chorioamnionitis, sepsis, postpartum endometritis, and postpartum wound infection. There is need to consider the foetus when offering antibiotic treatment to pregnant women is influenced to minimize potential adverse effects on the fetus. The choice of antibiotic treatment in postpartum infections is influenced by penetration into the breast milk. Penicillin or ampicillin work synergistically with gentamicin as the 1st line treatment of chorioamnionitis and endometritis. For patients with penicillin allergy, a 2nd or 3rd generation cephalosporin may be appropriate depending on the type of allergy, also clindamycin, erythromycin or vancomycin plus gentamicin. The treatment is given for 10 days. It is important to check for concurrent infections as other microorganism are often isolated **(40, 41)**.

4.2.3 Treatment of GBS infection in non-pregnant adults
Benzylpenicillin or ampicillin is the first-line treatment for sepsis, cellulitis, septic arthritis and meningitis. A second or third generation cephalosporin or a macrolide, or clindamycin may be appropriate for patients allergic to penicillin. It is advisable to seek specialist microbial advice as some cases may be complicated. Gentamicin may be considered as adjunctive therapy in such cases as presentation is rare and management complicated. Antibiotic treatment should be given for at least 10 days for sepsis and cellulitis; and 14 to 21 days for meningitis; and 3 to 4 weeks for septic arthritis **(40, 41)**. In case of osteomyelitis, conjunctivitis, otitis media, sinusitis, endocarditis, soft-tissue infection, and intra-abdominal infections, it may be necessary to get further microbial analysis in case infections are polymicrobial. As GBS infection occurs more commonly in older patients, hepatic and renal impairment must be taken into account when selecting dose of some drugs **(40)**. In case of urinary tract infections, the first-line treatment is amoxicillin or ampicillin. Penicillin or ampicillin with gentamicin is used for complicated infections such as pyelonephritis. Penicillin allergic patients should be given cotrimoxazole or nitrofurantoin, or

vancomycin with or without gentamicin. The duration of treatment course is 3 to 14 days (41).

For pneumonia, 1st line treatment is benzylpenicillin or ampicillin. Patients with penicillin allergy, a 2nd or 3rd generation cephalosporin may be appropriate depending on the type of allergy. Vancomycin, linezolid, a macrolide, or a quinolone may also be used but antibiotic susceptibility testing must be done before any treatment. Rifampicin or gentamicin may be considered as adjunctive therapy in selected cases (40, 41).

4.3 GBS and antibiotic resistance

Group B streptococcus remains susceptible to penicillin, the first-line agent for both prophylaxis and treatment of GBS infection (42). Until recently, patients with history of beta-lactam antibiotic allergy, clindamycin and erythromycin were recommended as alternatives. Unfortunately, resistance to these two antibiotics has been remarkably increasing (43-45).

High resistance rates to erythromycin and clindamycin have been observed in some studies (51% to erythromycin (46) and 54% to clindamycin (45). Because of possible resistance problems with erythromycin and clindamycin, vancomycin can be used as the initial treatment of GBS infection in patients who are allergic to penicillin (47).

Fluoroquinolones, especially the later generations, are active against GBS infections. Third and fourth generation fluoroquinolones are often recommended for treatment of pneumonia due to *S. agalactiae* (48). However, resistance to fluoroquinolones has been detected, even to levofloxacin, a third-generation fluoroquinolone (49-51).

4.3.1 Mechanisms of Group B Streptococcal antibiotic resistance

GBS uses different mechanisms to achieve antibiotic resistance: alteration of penicillin-binding protein by point mutation in resistance to beta-lactam antibiotics (52), acquisition of new genes coding for ribosome protection or efflux pump (53-55), acquisition of antibiotic resistance chromosomal transposon (55), chromosomal mutation (50), and plasmid-mediated resistance (56) (Table 4.2).

Beta-lactam antibiotics are bactericidal agents that act by inhibiting the peptidoglycan synthesis in bacteria cell wall (57). Although *S. agalactiae* remains susceptible to penicillin, ampicillin and first-generation cephalosporins, cases of high MICs in penicillin, ampicillin and cefazolin have been reported with alteration in penicillin-binding proteins **(4)**. Point mutations in *bpb2x, a* gene encoding a region of penicillin-binding protein were identified among GBS strains with elevated MICs to beta-lactam antibiotics tested **(52, 58)**.

Erythromycin and other macrolides prevent bacterial growth by binding to the 23rRNA molecule in the 50S subunit of the bacteria ribosome, blocking the exit of peptide chain **(57)**. Two common resistance mechanisms to erythromycin identified in GBS are methylation of 23S rRNA encoded by *erm* (erythromycin ribosome methylation) genes and drug efflux encoded by *mef* (macrolide efflux) genes **(53, 55)**.

As is the case with erythromycin, clindamycin and other lincosamides work mostly by binding to the 50S ribosomal subunit of bacteria, thereby interfering with the transpeptidation reaction and disrupting protein synthesis **(57)**. Resistance to clindamycin is also caused by acquisition of *erm* genes **(53, 55)**. The *inu*(B) gene was characterized and identified to be responsible for lincosamide nucleotidylation in GBS resistant isolates **(54)**.

Fluoroquinolones inhibit DNA synthesis through cleavage of bacterial DNA in gyrase and topoisomerase genes **(59)**. The mechanism of resistance in fluoroquinolones among *S. agalactiae* is due to target site alterations in the fluoroquinolone-resistance determining region (QRDR) of gyrase (*gyr*A , *gyr*B) and topoisomerase genes (*par*C, *par*E) **(49-51)**. The two enzymes are important in the replication, transcription, recombination and repair of DNA **(60)**. The *gyr*A mutations often occur at amino acid positions 81 and 85 and the *par*C mutations often occur at amino acid positions 79 and 83 **(61)**. Efflux-mediated resistance to fluoroquinolones has been demonstrated in *Staphylococcus aureus* and other streptococci **(62-64)**.

4.3.2 GBS antibiotic resistance and serotypes

Distribution of GBS serotypes among GBS resistant strains has been analyzed in many studies. Some serotypes are more likely to be resistant to antibiotics than the others, which might relate to better adaptability. In the United States, serotype V strains are more likely than other serotypes to be resistant to erythromycin and clindamycin **(43)**. Three quarters of serotype V can be resistant to clindamycin or erythromycin or both **(65)**. Serotype V is often dominant among GBS isolates with macrolide resistance in pregnant women **(66)**. In clinical samples most erythromycin-resistant GBS strains were also of serotype V **(67)**.

4.3.3 GBS antibiotic resistance among colonizing and invasive

Group B Streptococcus is found in the human body as part of the normal flora or as a pathogen. If it is not a pathogen, GBS then develops resistance to antibiotics taken by the individual. Difference in the prevalence of resistance to antibiotics between colonizing and invasive GBS isolates has not been clearly demonstrated. However resistance to erythromycin and clindamycin has been recorded to be higher among colonizing strains compared to invasive ones (7, 68). More research is needed to identify any potential differences in antimicrobial resistance between nonpathogenic and pathogenic GBS.

Table 4.2 Mechanisms of resistance of Group B Streptococcal resistance to major antibiotic groups.

Antibiotic	Mechanisms of resistance
Beta-lactams	Point mutation in *pbp2x*
Fluoroquinolones	Mutations in gyrase (*gyr*A, *gyr*B) and topoisomerase (*par*C, *par*E)
Erythromycin	Ribosomal methylase (acquisition of *erm*(A), *erm*(B), *erm*(C), *erm*(TR) genes) Efflux pump (acquisition of *mef*(A), *mef*(E) genes)
Tetracycline	Ribosome protection (acquisition of *tet*(M), *tet*(O), *tet* (S), *tet* (T) genes) Efflux by proton antiporters (acquisition of *tet* (L), *tet* (K) genes) Plasmid-mediated
Streptomycin Kanamycin Gentamicin	Acquisition of *aphA*-3 gene Acquisition of chromosomal gentamicin resistance transposon Tn3706
Lincosamides	Ribosomal methylase (acquisition of *erm*(A), *erm*(B), *erm*(C), *erm*(TR) genes) Lincosamide nucleotidylation (acquisition of *lnu* (B) gene)

4.4 Main references

1. **Pettersson, K.** 2007. Perinatal infection with Group B streptococci. Semin Fetal Neonatal Med **12**:193-7.

2. **Verani, J. R., L. McGee, and S. J. Schrag.** 2010. Prevention of perinatal group B streptococcal disease--revised guidelines from CDC, 2010. MMWR Recomm Rep **59**:1-36.

3. **Schrag, S. J., E. R. Zell, R. Lynfield, A. Roome, K. E. Arnold, A. S. Craig, L. H. Harrison, A. Reingold, K. Stefonek, G. Smith, M. Gamble, and A. Schuchat.** 2002. A population-based comparison of strategies to prevent early-onset group B streptococcal disease in neonates. N Engl J Med **347**:233-9.

4. CDC. Prevention of perinatal group B streptococcal disease: revised guidelines from CDC. MMWR 2002;**51**(No. RR-11).

5. **Trijbels-Smeulders M., L. Kolée, A. H. Adriaanse, J. H. Kimpen, and L. J. Gerads.** 2004. Neonatal group B streptococcal infection: Incidence and strategies for prevention in Europe. Pediatr Infect Dis J **23**:172-173.

6. **Colburn T, and R. Gilbert.** 2007. An overview of the natural history of early onset group B streptococcal disease in the UK. Early Hum Dev **83**:149-156.

7. **Borchardt, S.M., J.H. DeBusscher, P.A. Tallman, S. D. Manning, C.F. Marrs, T.A. Kurzynski, and B. Foxman.** 2006. Frequency of antimicrobial resistance among invasive and colonizing Group B Streptococcal isolates. BMC Infectious Diseases. **6**:57.

8. **Fluegge K., S. Supper, A. Siedler, and R. Berner.** 2004. Antibiotic susceptibility in neonatal invasive isolates of *Streptococcus agalactiae* in a 2-year nationwide surveillance study in Germany. Antimicrob Agents Chemother **48**:4444-4446.

9. **Lancefield R. C.** 1934. Serological differentiation of specific types of bovine haemolytic streptococci (group B). Exp Med **59**:441-458.

10. **Baker C. J., M. A. Rench, M. S. Edwards, R. J. Carpenter, B. M. Hays, and D. L. Kasper.** 1988. Immunization of pregnant women with a

polysaccharide vaccine of group B streptococcus. N Engl J Med **319**:1180-1185.

11. **Kotloff K. L., A. Fattom, L. Basham, A. Hawwari, S. Harkonen, and R. Edelman.** 1996. Safety and immunogenicity of a tetravalent group B streptococcal polysaccharide vaccine in healthy adults. Vaccine **14**:446-450.

12. **Lagergård T., J. Shiloach, J. B. Robbins, and R. Schneerson.** 1990. Synthesis and immunological properties of conjugates composed of group B streptococcus type III capsular polysaccharide covalently bound to tetanus toxoid. Infect Immun **58**: 687-694.

13. **Paoletti L. C., M. R. Wessels, A. K. Rodewald, A. A. Shroff, H. L. Jennings and D. L. Kasper.** 1994. Neonatal mouse protection against infection with multiple group B streptococcal (GBS) serotypes by immunization with a tetravalent GBS polysaccharide-tetanus toxoid conjugate vaccine. Infect Immun; **62**:3236-3243.

14. **Baker C. J., M. A. Rench, M. Fernandez, L. C. Paoletti, D. L. Kasper, and M. S. Edwards.** 2003. Safety and immunogenicity of a bivalent group B streptococcal conjugate vaccine for serotypes II and III. J Infect Dis; **188**:66-73.

15. **Baker C. J., M. Rench, and P. McInnes.** 2003. Immunization of pregnant women with group B streptococcal type III capsular polysaccharide-tetanus toxoid conjugate vaccine. Vaccine **21**:4368-4372.

16. **Madoff, L. C., J. L. Michel, E. W. Gong, A. K. Rodewald, and D. L. Kasper.** 1992. Protection of neonatal mice from group B streptococcal infection by maternal immunization with beta C protein. Infect Immun **60**:4989-94.

17. **Larsson C., M. Stålhammar-Cedermalm, and G. Lindahl.** 1996. Experimental vaccination against group B streptococcus, an encapsulated bacterium with highly purified preparations of cell surface proteins Rib and alpha. Infect Immun; **64**:3518-3523.

18. **Lancefield, R. C.** 1938. Two serological types of group B hemolytic
66

streptococci with related, but not identical, type specific substances. J Exp Med **67**:25-40.

19. **Baker, C. J., and D. L. Kasper.** 1976. Correlation of maternal antibody deficiency with susceptibility to neonatal group B streptococcal infection. N Engl J Med **294**:753-6.

20. **Paoletti, L. C., and D. L. Kasper.** 2002. Conjugate vaccines against group B Streptococcus types IV and VII. J Infect Dis **186**:123-6.

21. **Paoletti, L. C., L. C. Madoff, and D. L. Kasper.** 2006. Surface structures of group B streptococcus important in human immunity, p. 169-185. *In* V. A. Fischetti, Novick, R. P., Ferretti, J. J., Portony, D. A.and Rood, J. I. (ed.), Gram-Positive Pathogens ASM Press, Washington DC.

22. **Zaleznik, D. F., M. A. Rench, S. Hillier, M. A. Krohn, R. Platt, M. L. Lee, A. E. Flores, P. Ferrieri, and C. J. Baker.** 2000. Invasive disease due to group B Streptococcus in pregnant women and neonates from diverse population groups. Clin Infect Dis **30**:276-81.

23. **Lachenauer, C. S., D. L. Kasper, J. Shimada, Y. Ichiman, H. Ohtsuka, M. Kaku, L. C. Paoletti, P. Ferrieri, and L. C. Madoff.** 1999. Serotypes VI and VIII predominate among group B streptococci isolated from pregnant Japanese women. J Infect Dis **179**:1030-3.

24. **Lindahl, G., M. Stalhammar-Carlemalm, and T. Areschoug.** 2005. Surface proteins of *Streptococcus agalactiae* and related proteins in other bacterial pathogens. Clin Microbiol Rev **18**:102-27.

25. **Brodeur, B. R., M. Boyer, I. Charlebois, J. Hamel, F. Couture, C. R. Rioux, and D. Martin.** 2000. Identification of group B streptococcal Sip protein, which elicits cross-protective immunity. Infect Immun **68**:5610-8.

26. **Gravekamp, C., D. L. Kasper, L. C. Paoletti, and L. C. Madoff.** 1999. Alpha C protein as a carrier for type III capsular polysaccharide and as a protective protein in group B streptococcal vaccines. Infect Immun **67**:2491-6.

27. **Yang, H. H., S. J. Mascuch, L. C. Madoff, and L. C. Paoletti.** 2008. Recombinant group B Streptococcus alpha-like protein 3 is an effective

immunogen and carrier protein. Clin Vaccine Immunol **15**:1035-41.

28. **Paoletti, L. C., and L. C. Madoff.** 2002. Vaccines to prevent neonatal GBS infection. Semin Neonatol **7**:315-23.

29. **Cheng, Q., B. Carlson, S. Pillai, R. Eby, L. Edwards, S. B. Olmsted, and P. Cleary.** 2001. Antibody against surface-bound C5a peptidase is opsonic and initiates macrophage killing of group B streptococci. Infect Immun **69**:2302-8.

30. **Santillan, D. A., M. E. Andracki, and S. K. Hunter.** 2008. Protective immunization in mice against group B streptococci using encapsulated C5a peptidase. Am J Obstet Gynecol **198**:114 e1-6.

31. **Heath, P. T., and R. G. Feldman.** 2005. Vaccination against group B streptococcus. Expert Rev Vaccines **4**:207-18.

32. **Seepersaud, R., S. B. Hanniffy, P. Mayne, P. Sizer, R. Le Page, and J. M. Wells.** 2005. Characterization of a novel leucine-rich repeat protein antigen from group B streptococci that elicits protective immunity. Infect Immun **73**:1671-83.

33. **Stalhammar-Carlemalm, M., J. Waldemarsson, E. Johnsson, T. Areschoug, and G. Lindahl.** 2007. Nonimmunodominant regions are effective as building blocks in a streptococcal fusion protein vaccine. Cell Host Microbe **2**:427-34.

34. **Gourlay, L. J., I. Santi, A. Pezzicoli, G. Grandi, M. Soriani, and M. Bolognesi.** 2009. Group B streptococcus pullulanase crystal structures in the context of a novel strategy for vaccine development. J Bacteriol **191**:3544-52.

35. **Margarit, I., C. D. Rinaudo, C. L. Galeotti, D. Maione, C. Ghezzo, E. Buttazzoni, R. Rosini, Y. Runci, M. Mora, S. Buccato, M. Pagani, E. Tresoldi, A. Berardi, R. Creti, C. J. Baker, J. L. Telford, and G. Grandi.** 2009. Preventing bacterial infections with pilus-based vaccines: the group B streptococcus paradigm. J Infect Dis **199**:108-15.

36. **Goldenberg, R. L., E. M. McClure, S. Saleem, D. Rouse, and S. Vermund.** 2006. Use of vaginally administered chlorhexidine during labor to improve pregnancy outcomes. Obstet Gynecol **107**:1139-46.

37. **Stade, B., V. Shah, and A. Ohlsson.** 2004. Vaginal chlorhexidine during labour to prevent early-onset neonatal group B streptococcal infection. Cochrane Database Syst Rev:CD003520.

38. **Florindo C.; Gomes J.P.; Rato M.G.; Bernardino L.; Spellerberg B.; Santos-Sanches I. and Borrego M.J.** 2011. Molecular epidemiology of group B Streptococcal meningitis in children beyond the neonatal period from Angola. J Med Microbiol 60(Pt 9):1276-80

39. **Simoes J. A., A. A Aroutcheva, I. Heimler, and S. Faro.** 2004. Antibiotic resistance patterns of group B streptococcal clinical isolates. Infect Dis Obstet Gynecol **12**:1-8.

40. **Woods C.J. and C. S. Levy.** 2010. GBS infections: Treatment and Medication.eMedicine-Medscape.http://eMedicine-medscape.com/article/GBS.

41. **Narayanan S. K., M. Ossiani, and C. S. Levy.** 2006. Streptococcus Group B Infections. eMedicine from WebMD. http://eMedicine-streptococcus-Group-B-infections. Clinical Reference. Page1-14.

42. **Phares, C. R., R. Lynfield, M.M. Farley, J. Mohle-Boetani, L.H. Harrison, S. Petit, A. S. Craig, W. Schaffner, S. M. Zansky, K. Gershman, K. R. Stefonek, B. A. Albanese, E. R. Zell, A. Schuchat, S. J. Schrag; Active Bacterial Core surveillance/Emerging Infections Program Network.** 2008. Epidemiology of invasive group B streptococcal disease in the United States,1999-2005. JAMA. **299**:2056-65.

43. **Castor, M.L., C. G. Whitney, K. Como-Sabetti, R. R. Facklam, P. Ferrieri, J. M. Bartkus, B. A. Juni, P. R. Cieslak, M. M. Farley, N. B. Dumas, S. J. Schrag, and R. Lynfield.** 2008. Antibiotic resistance patterns in invasive group B streptococcal isolates. Infect Dis Obstet Gynecol. **2008**:727505.

44. **Gygax, S. E., J. A. Schuyler, L. E. Kimmel, J. P. Trama, E. Mordechai, and M. E. Adelson.** 2006. Erythromycin and clindamycin resistance in Group B Streptococcal clinical isolates. Antimicrob Agents Chemother **50**:1875–1877.

45. Janapatla, R. P., Y. R. Ho, J. J. Yan, H. M. Wu, J. J. Wu. 2008. The prevalence of erythromycin resistance in group B streptococcal isolates at a University Hospital in Taiwan. Microb Drug Resist **14**:293-297.

46. Back, E. E., E. J. O'Grady, and J. D. Back. 2012. High rates of perinatal Group B *Streptococcus* clindamycin and erythromycin resistance in an Upstate New York Hospital. Antimicrob Agents Chemother **56**:739-742.

47. Gilbert, D. N., R.C. Moellering, G. M. Eliopoulos, H.F. Chambers, and M.S. Saag. 2011. The Sanford Guide to Antimicrobial Therapy. Antimicrobial Therapy Inc.

48. Blasi, F., S. Ewig, A. Torres, and G. Huchon. 2006. A review of guidelines for antibacterial use in acute exacerbations of chronic bronchitis. Pulm Pharmacol Ther **19**:361-9.

49. Savoia, D., C. Gottimer, C. Crocilla, and M. Zucca. 2008. *Streptococcus agalactiae* in pregnant women: phenotypic and genotypic characters. J Infect. **56**:120-125. 128.

50. Wehbeh, W., R. Rojas-Diaz, X. Li, N. Mariano, L. Grenner, S. Segal-Maurer, B. Tommasulo, K. Drlica, C. Urban, and J. J. Rahal. 2005. Fluoroquinolone-Resistant *Streptococcus agalactiae*: epidemiology and mechanism of resistance. Antimicrob Agents Chemother **49**:2495-2497.

51. Wu, H. M. , R. P. Janapatla, Y-R Ho, K-H Hung, C-W Wu, J-J Yan, and J-J Wu. 2008. Emergence of fluoroquinolone resistance in Group B Streptococcal isolates in Taiwan. Antimicrob Agents Chemother **52**:1888-1890.

52. Dahesh, S., M. E. Hensler, N. M. Van Sorge, R. E. Gertz Jr., S. Schrag, V. Nizet, and B. W. Beall. 2008. Point Mutation in the Group B Streptococcal *pbp2x* gene conferring decreased susceptibility to β -Lactam antibiotics. Antimicrob Agents Chemother **52**:2915–2918.

53. Heelan, J. S., M. E. Hasenbein, and A. J. McAdam. 2004. Resistance of Group B *Streptococcus* to selected antibiotics, including erythromycin and clindamycin. J Clin Microbiol **42**:1263–1264.

54. **Malbruny, B., A. M. Werno, T. P. Anderson, D. R. Murdoch, and R. Leclercq.** 2004. A new phenotype of resistance to lincosamide and streptogramin A-type antibiotics in *Streptococcus agalactiae* in New Zealand. J Antimicrob Chemother **54**:1040–1044.

55. **Poyart, C. L. Jardy, G. Quesne, P. Berche, and P. Trieu-Cuot.** 2003. Genetic basis of antibiotic resistance in *Streptococcus agalactiae* strains isolated in a French Hospital. Antimicrob Agents Chemother **47**:794-797.

56. **Burdett, V.** 1980. Identification of tetracyclin-resistant-R-plasmids in *Streptococcus agalactiae*. Antimicrob Agents Chemother **18**:753–760.

57. **Tenover, F.C.** 2006. Mechanisms of antimicrobial resistance in bacteria. American J Med **119**:S3-S10.

58. **Kimura, K., J. Wachino, S. Kurokawa, K. Suzuki, N. Yamane, Y. Shibata, and Y. Arakawa.** 2006. Emergence of penicillin-resistant group B streptococci. Abstr. 46th Intersci. Conf. Antimicrob. Agents Chemother., abstr. C2-1286.

59. **Oliphant, C. M., and G. M. Green.** 2002. Quinolones: A comprehensive review. Am Fam Physician **65**:455-465.

60. **Jacoby, G. A.** 2005. Mechanisms of resistance to quinolones. Clin Infect Dis. **15**:S120-6.

61. **Biedenbach, D. J., M. A. Toleman, T. R. Walsh, and R. N. Jones.** 2006. Characterization of fluoroquinolone-resistant beta-hemolytic *Streptococcus* spp. isolated in North America and Europe including the first report of fluoroquinolone-resistant *Streptococcus dysgalactiae* subspecies *equisimilis*: report from the SENTRY Antimicrobial Surveillance Program (1997-2004). Diagn Microbiol Infect Dis **55**:119-27.

62. **Jumbe, N., A. Louie, W. Miller, W. Liu, M. Deziel, V. Tam, R. Bachhawat, and G. Drusano.** 2006. Quinolone efflux pumps play a central role in emergence of fluoroquinolone resistance in *Streptococcus pneumoniae*. Antimicrob Agents Chemother **50**:310-317.

63. **Kaatz, G. W., S. M. Seo, and C. A. Ruble.** 1993. Efflux-mediated

fluoroquinolone resistance in *Staphylococcus aureus*. Antimicrob Agents Chemother **37**:1086–1094.

64.**Piddock, L. J. V.** 2006. Clinically relevant chromosomally encoded multidrug resistance efflux pumps in bacteria. Clin Microbiol Rev **19**:382-402.

65.**Lee, B. K., Y. R. Song, M. Y. Kim, J. H. Yang, J. H. Shin, Y. S. Seo, K. Y. Oh, H. R. Yoon, S. Y. Pai, B. Foxman, and M. Ki.** 2010. Epidemiology of group B *Streptococcus* in Korean pregnant women. Epidemiol Infect **138**:292-298.

66.**Brzychczy-Włoch, M., T. Gosiewski, M. Bodaszewska, W. Pabian, M. Bulanda , P. Kochan, M. Strus and P.B. Heczko.** 2010. Genetic characterization and diversity of *Streptococcus agalactiae* isolates with macrolide Resistance. J Med Microbiol **59**:780–786.

67.**De Francesco, M. A., S. Caracciolo, F. Gargiulo, and N. Manca.** 2012. Phenotypes, genotypes, serotypes and molecular epidemiology of erythromycin-resistant *Streptococcus agalactiae* in Italy. Eur J Clin Microbiol Infect Dis. **31(8)**:1741-7.

68.**De Azavedo, J. C. S., M. Mcgavin, C. Duncan, D. E. Low, and A. Mcgeer.** 2001. Prevalence and mechanisms of macrolide resistance in invasive and noninvasive Group B *Streptococcus* isolates from Ontario, Canada. Antimicrob Agents Chemother **45**: 3504–3508.

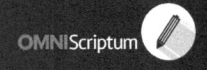

Printed by Books on Demand GmbH, Norderstedt / Germany